The Essential Guide to

Precalculus

유하림(Harim Yoo) 지음

HERMONHOUSE

Preface

To. 학부모님과 학생들께

올해 봄에 출판한 The Essential Guide to Competition Math(Fundamentals) 이후 바로 작업한 The Essential Guide to Precalculus를 이번 여름에 출판하게 되었습니다. 기존 교과서 및 교재들의 단점이라고 느껴졌던 점은 기본유형에 대한 문제만 다루고 있다는 점이었습니다. 조금 더 구체적으로 말하자면, AMC 12 및 다양한 경시대회까지 도전하고자 하는 많은 유학생들에게 삼각함수의 기본유형만 가지고 준비를 하기에는 항상 어려움이 있다는 것을 현장에서 뼈저리게 느끼고 있던 터였습니다.

기존에 출판한 4권의 교과대비 교재의 정점을 찍는 Precalculus 교재를 출판하며, 기대가 매우 큽니다. 이미 유수의 대학교에 입학한 제자들에게 여러 차례 확인 및 검수를 받으면서 들었던 연구원들의 피드백 중 본인이 "10학년 때 이 교재가 있으면 좋았었겠다"라는 내용과 "개념을 깊이 생각하고, 적용해야 풀리는 문제들 위주로 작성되었다"라는 내용이 가장 인상적이었습니다.

처음으로 Precalculus를 배우는 학생들부터, 상위권 학생들까지 다 아우를 수 있는 내용뿐만 아니라, 위 개념이 누구에 의해서 어떻게 발전되었는지, 어떻게 우리 사회에 영향을 주었는지까지 역사적 관점에서도 서술하였는데, 이 부분이 실제로 수업을 하는 학생들에게 효과가 매우 컸습니다. 이로 인해, 앞으로 교재의 방향을 잡으면서 교재 작업을 해온 터라, 이번 교재 출판은 제게 큰 의미가 있습니다.

이 교재를 출간할 수 있도록 물심양면으로 힘써주신 마스터프렙 권주근 대표님께 감사합니다. 또한, 제게 항상 롤모델이 되어주고, 강사로서 성장할 수 있는 원동력을 주시고 계신 심현성 대표님께도 감사의 마음을 전합니다. 그리고 이 책을 완성할 수 있도록 제 기초를 만들어주신 제 아버지, 어머니께 고맙습니다. 언제나 든든한 지원군인 제 아내와 딸에게도 고맙다는 말을 전달합니다. 마지막으로, 제 삶에 이러한 기회를 주신 하나님께 감사드립니다. 앞으로도 더 좋은 교재를 만들어 견고하고 튼튼한 유하림 커리큘럼을 준비하도록 하겠습니다.

2021년 여름
유하림

저자 소개

유하림(Harim Yoo)

미국 Northwestern University,
B.A. in Mathematics and Economics
(노스웨스턴 대학교 수학과/경제학과 졸업)

마스터프렙 수학영역 대표강사
압구정 현장강의 ReachPrep 원장

고등학교 시절 문과였다가, 미국 노스웨스턴 대학교 학부 시절 재학 중 수학에 매료되어, Calculus 및 Multivariable Calculus 조교 활동 및 수학 강의 활동을 해온 문/이과를 아우르는 독특한 이력을 가진 강사이다. 현재 압구정 미국수학/과학전문 학원으로 ReachPrep(리치프렙)을 운영 중이며, 미국 명문 보딩스쿨 학생들과 국내 외국인학교 및 국제학교 학생들을 꾸준히 지도하면서 명성을 쌓아가고 있다.

2010년 자기주도학습서인 "몰입공부"를 집필한 이후, 미국 중고교수학에 관심을 본격적으로 가지게 되었고, 현재 유하림커리큘럼 Essential Math Series를 집필하여, 압구정 현장강의 미국 수학프리패스를 통해, 압도적으로 많은 학생들의 피드백을 통해, 발전적으로 교재 집필에 힘쓰고 있다.

마스터프렙 수학영역 대표강사 중 한 명으로 미국 수학 커리큘럼의 기초수학부터 경시수학까지 모두 영어와 한국어로 강의하면서, 실전 경험을 쌓아 그 전문성을 확고히 하고 있다.

[저 서] 몰입공부 (랜덤하우스코리아)
The Essential Workbook for SAT Math Level 2 (이담북스)
The Essential Guide to Prealgebra (헤르몬하우스)
The Essential Guide to Algebra 1 (헤르몬하우스)
The Essential Guide to Geometry (헤르몬하우스)
The Essential Guide to Algebra 2 (헤르몬하우스)
The Essential Guide to Precalculus (헤르몬하우스)
The Essential Guide to Competition Math (Fundamentals) (헤르몬하우스)
The Essential Guide to Competition Math (Number Theory) (헤르몬하우스)
The Essential Guide to Competition Math (Counting and Probability) (헤르몬하우스)
The Essential Workbook for Geometry (헤르몬하우스)

저자직강 인터넷 강의는 SAT, AP No.1 인터넷 강의 사이트인 마스터프렙 (www.masterprep.net) 에서 보실 수 있습니다.

이 책의 특징

유하림 커리큘럼 Essential Math Series 교과 과목 대비를 위한 책 중 방점을 찍는 교재로, Precalculus 과목을 준비하는 교재입니다. 위 과목을 처음 공부하는 학생부터, AMC 12 및 그 이상의 경시대회에 나오는 주제들까지 다뤄보고 싶은 학생들을 위한 필독서가 되길 바라는 마음으로 집필하였습니다. 현재 미국 명문 Boarding School 및 국내외 외국인학교에 다니는 9, 10학년 학생들이 학교 시험뿐 아니라, AMC 12 및 HMMT와 같은 타 경시대회에서 삼각함수 관련 주제에 관한 생각을 실제로 적용하고, 문제 풀이의 방향을 잡을 수 있길 바라면서 책을 썼습니다.

1st
Precalculus에 몰입할 수 있는 책

경시 수학 관련하여 압구정 현장 강의에서 많은 학생들을 교육하다보니, 해당 개념들의 최초 발견부터 해서 발전까지 궁금해하는 학생들이 많았습니다. 이로 인해, 이전까지 없던 교재를 만들고자 하였는데, 특히 수학 역사가 융합된 교재를 만들게 되었습니다. 이 교재로 현장 강의를 통해 수업하다 보면, 처음 배우는 학생들은 매우 신선하게 받아들이며, 상위권 학생들에게 자극을 줄 수 있는 고난이도 문제들까지 포함되어있어 두루두루 다양한 학생들이 즐겁게 배우는 것을 알 수 있었습니다. 기존에 나온 교재들의 단순한 개념 설명보다, 확실하게 "몰입"할 수 있는 형태로 서술되어 있어, 인도, 아랍, 그리고 유럽에 걸쳐 발전되어 온 삼각함수의 응용과 그 형태들을 재미있게 배울 수 있습니다.

2nd
생각의 확장을 위한 책

Precalculus의 개념을 배우고, 그 개념의 폭좁은 응용을 공부하는 것이 아니라, 조금은 깊이 생각해보고, 유수의 수학자들이 위 개념들을 처음 접했을 때 고민했던 문제들을 포함하여, 이 과목을 준비하는 학생들이 단순히 개념을 적용하는 수준에서 머무는 것이 아니라, 진지하게 "생각" 할 수 있는 수준의 교재로 집필하였습니다. 인터넷에서 조금만 찾아보면 나오는 답들로 구성된 교재가 아닌, 혼자서 끙끙대며 고민해보면서 풀어봐야 그 해답을 찾을 수 있는 문제들 위주로 구성되어 있어, 가르치는 선생에게도 배우는 학생에게도 도전을 주는 교재입니다.

3rd
유학 준비생을 위한 바로 그 책

교과 수학보다 응용의 폭이 깊어서, 시작조차 엄두를 내지 못했던 유학생들과 그 준비생들에게 하나의 지름길이 될 수 있기를 희망하면서 집필한 책입니다. 노스웨스턴 대학교 학창시절 수학에 대한 열정을 뒤늦게 꽃피워 밤새워 공부했고, 저는 학생들을 더 잘 가르치고, 더 나은 미래로 이끌기 위해, AMC, AIME, ARML, HMMT, PUMaC, SUMO와 같은 문제들을 동일한 열정으로 끊임없이 풀고 해석합니다. 여러분이 지금 보는 이 책은 제 현재 노력의 최선의 산실이며, 앞으로도 그러할 것입니다. 이 책을 통해 수학을 두려워하지 않고, 문제 해결을 즐거워하며, 이른 나이에 수학에 대한 열정을 꽃피우길 기대합니다.

CONTENTS

Solution Manual

TOPIC

1

Essence of Graphs

The position of a point in a plane is given by an ordered pair (x, y). This coordinate system is widely known by Rene Descartes, but it was invented by a bishop mathematician Nicole Oresme. It measures the displacement of (x, y) from two perpendicular axes whose intersection point is the origin $(0, 0)$.

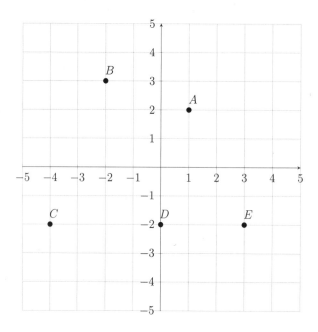

Given a point (x, y), a x-coordinate is sometimes called *abscissa*, and y-coordinate is called the *ordinate*. In the above case, the coordinates of A are $(1, 2)$, and it is written as $A(1, 2)$. Points could also lie on either axis[1]. For instance, the coordinates of D are $(0, -2)$.

1. Let $A(1, 2)$, $B(-2, 3)$, and $D(0, -2)$. Find the area of a triangle $\triangle ABD$, simply using the typical area formula, i.e, $\frac{1}{2} \times$ base $\times$ height.

---

[1]The y-axis or the x-axis are not defined as any part of the quadrants.

As you can see from the diagram above, when we plot any point, we use the coordinate system. For instance, suppose you are given with (a, b). Then, we plot the point by moving a units horizontally rightwards from the origin and b units vertically above from the origin to plot the point. Specifically, suppose you have a point $(1, 2)$. We start from the Origin, which is the intersection point of the x-axis (the thick horizontal line) and the y-axis (the thick vertical line). Then, we move 1 unit right and 2 unit up from $(0, 0)$. The following formulas are both midpoint formula and distance formula in two dimensional cases.

- Midpoint formula[2] : given two points $A(a, b)$ and $B(c, d)$, the midpoint of $\overline{AB}$ is

$$\left(\frac{a + c}{2}, \frac{b + d}{2} \right)$$

- Distance formula[3] : given two points $A(a, b)$ and $B(c, d)$, the distance between A and B is

$$\sqrt{(a - c)^2 + (b - d)^2}$$

The midpoint of $(1, 4)$ and $(3, 6)$ can be found by separate caseworks. The midpoint of 1 and 3 must be 2, and that of 4 and 6 is 5. Hence, the midpoint of the two points is $(2, 5)$. As you can see from the example, we get the midpoints of the x-coordinates and y-coordinates, separately. When we get the coordinates, we can simply compute in terms of one dimensional case.

2. Suppose A, B, and C are collinear. If $AB : BC = 2 : 3$, where $A = (1, 4)$ and $C = (6, -1)$, find the coordinates of B.

---

[2]Try to understand this in terms of A, B, and M, all collinear points, where M is cutting $\overline{AB}$ into half. This is crucial in understanding collinear points in vector section.

[3]There are two types of distance formula-one dimensional case and two dimensional case. If either x-coordinate or y-coordinate is fixed, then we simply find the difference of the other coordinates. On the other hand, if any coordinate is distinct, then we use Pythagorean theorem, as stated in the bullet point.

Out of symmetries, three major symmetries are axis symmetries and the origin symmetry.

- If a graph is symmetric with respect to the x-axis, and there is a point (x, y) on the graph, then a point $(x, -y)$ is also on the graph.

- If a graph is symmetric with respect to the y-axis, and there is a point (x, y) on the graph, then a point $(-x, y)$ is also on the graph.

- If a graph is symmetric with respect to the origin, and there is a point (x, y) on the graph, then a point $(-x, -y)$ is also on the graph.

3. Find the point that satisfies the following conditions.

(a) $(2, 3)$ reflected about the x-axis.

(b) $(3, 2)$ reflected about the line $y = 1$.

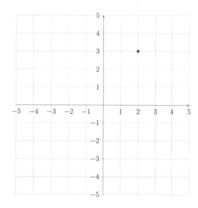 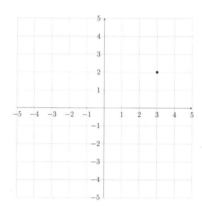

That being said, we should be able to check whether the given function is really symmetric, without referring to a graphic calculator.

- x-axis : Replace y by $-y$ and see if the resulting equation is equivalent to the original equation.

- y-axis : Replace x by $-x$ and see if the resulting equation is equivalent to the original equation.

- Origin : Replace both x, y by $-x, -y$ and see if the resulting equation is equivalent to the original equation.

Example

Determine whether $f(x) = x^3 + 3x^2$ is symmetric with respect to the y-axis.

Solution
In order to check the symmetry with respect to the y-axis, we need to put $-$ in front of x. $f(-x) = (-x)^3 + 3(-x)^2 = -x^3 + 3x^2 \neq x^3 + 3x^2 = f(x)$. This shows that if we put $(-)$ in front of x, the resulting expression is different from the original one, so it cannot be symmetric.

4. Which of the following functions satisfies $f(-x) = f(x)$?

(A) $f(x) = x + 3x^2$
(B) $f(x) = 3 - 2x$
(C) $f(x) = |x| + 2$
(D) $f(x) = 2x + 3$
(E) $f(x) = 2(x + 1)$

5. Reflect the following portions of the graph

(a) about the origin.

(b) about the y-axis.

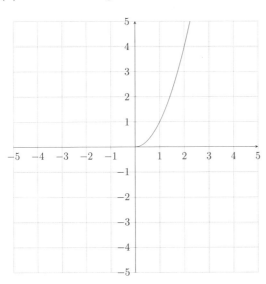

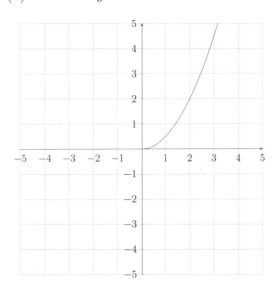

Why are these symmetric properties useful in graphing? The answer to this may be related to the existence of the point of symmetry or line of symmetry. Normally, if the graph is symmetric about geometric objects, either point or line, it is easy for us to visualize(or guess) how the graph looks like in overall behavior.

- Symmetric about a point : graph the northeast portion of the graph and rotate the image counterclockwise about 180°.

- Symmetric about a line : graph the east portion of the graph and reflect it about the line of symmetry.

The strategies written above may look simple, but it reduces the amount of work for us to predict how a graph looks like.

6. If $(2, 4)$ and $(-4, 0)$ are symmetric with respect to the point (a, b), find the values of a and b.

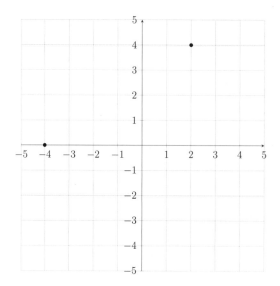

If we have a function whose graph is symmetric with respect to the y-axis or the origin, we give them special names.

- Even function : a function $f(x)$ that satisfies $f(-x) = f(x)$.

- Odd function : a function $f(x)$ that satisfies $f(-x) = -f(x)$.

7. If a graph is symmetric with respect to the y-axis, and the point $(2, 3)$ is on the graph, then which of the following points must be on the graph?

(A) $(2, -3)$ (B) $(-2, 3)$ (C) $(-2, -3)$ (D) $(3, 2)$ (E) $(-3, -2)$

8. Test $y = \dfrac{x^2}{3x^2 + 5|x|}$ for symmetry.

(a) *Line Symmetry* : Is the graph symmetric about the y-axis?

(b) *Line Symmetry* : Is the graph symmetric about the x-axis?

(c) *Point Symmetry* : Is the graph symmetric about the origin?

9. Given an equation $\dfrac{x^2}{4} + \dfrac{y^2}{9} = 1$, which of the following line symmetries is true?

(A) The graph is symmetric only about the line $x = 0$.
(B) The graph is symmetric only about the line $y = 0$.
(C) The graph is symmetric only about the lines $x = 0$ and $y = 0$.
(D) The graph is symmetric only about the line $y = x$.
(E) The graph is symmetric only about the line $y = -x$.

Circle is a set of points equidistant from a fixed point called the center. In other words,

$$(x-a)^2 + (y-b)^2 = r^2$$

where (a, b) is the fixed point, called the center, and r is its radius.

The easiest type of problems related to circles includes **finding its center and radius**. Completing the square is the way to handle this question. Look at the following example.

If $x^2 + 4x + y^2 - 6y = 3$, then you may convert it using $x^2 + 4x + 4 + y^2 - 6y + 9 = 3 + 4 + 9 = 16$, which turns into $(x+2)^2 + (y-3)^2 = 4^2$.

A bit more advanced type of questions includes **the relationship between a line and circle**, which involves a tangent line. In order to solve the question, we either solve the equation or find the discriminant.

For instance, if $y = x + k$ and $y = x^2 + 3x + 2$ are tangent, set the two equations equal and solve them by discriminant, using that the distance between a line and the center of a circle is r.

$$x^2 + 3x + 2 = x + k$$
$$x^2 + x + (2 - k) = 0$$

Here, if you find the k-value satisfying $b^2 - 4ac = 1 - 4(2 - k) = 0$, the line and the curve are tangent.

Intermediate type of questions involving circle relates a circle equation passing through two points with the perpendicular bisector. Have a close look at the following figure. If A and B are the points on the circle, the perpendicular bisector of A and B must pass through the center of the circle.

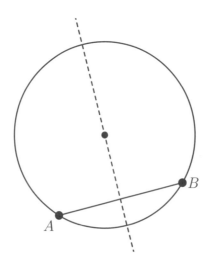

The points of intersection between two circles can be found by algebraic method, instead of graphing them. For example, if $(x-1)^2 + (y-1)^2 = 1$ and $x^2 + y^2 = 1$ are given, let them equal.

$$(x-1)^2 + (y-1)^2 = x^2 + y^2$$
$$x^2 - 2x + 1 + y^2 - 2y + 1 = x^2 + y^2$$
$$2 = 2x + 2y$$
$$1 = x + y$$
$$-x + 1 = y$$

The points of intersection must pass through the line $y = -x + 1$. Needless to say, if circle equation turns into circle inequality, then we must be able to interpret its meaning properly.

- $(x-h)^2 + (y-k)^2 < r^2 : (x, y)$ is inside the circle.

- $(x-h)^2 + (y-k)^2 > r^2 : (x, y)$ is outside the circle.

10. Which of the following is an interior point of a circle centered at the origin with the radius of 5?

(A) $(4, 4)$
(B) $(-4, 3)$
(C) $(3, -4)$
(D) $(2, 3)$
(E) $(-3, -4)$

11. The area of a circle $x^2 + y^2 - 10y = 144$ is

(A) 144π
(B) 169π
(C) 196π
(D) 225π
(E) 256π

12. Which of the following equations is the circle whose center is at the origin and tangent to the line equation $4x + 3y = 10$?

(A) $x^2 + y^2 = 2$
(B) $x^2 + y^2 = 4$
(C) $x^2 + y^2 = 3$
(D) $x^2 + y^2 = 5$
(E) $x^2 + y^2 = 10$

13. If the circle passes through $(2, 1)$ and $(6, 5)$, which of the following can be the center of the circle?

(A) $(3, 3)$
(B) $(2, 4)$
(C) $(3, 5)$
(D) $(2, 5)$
(E) $(5, 1)$

Here is the definition of a curve known as parabola. **Parabola** is the set of points whose distance to a fixed point is equal to the distance to the given line. If we call the signed distance between the vertex and the focus as p, then two parabolas opening either left-right or open up-down can be given by

$$y^2 = 4px \qquad\qquad x^2 = 4py$$

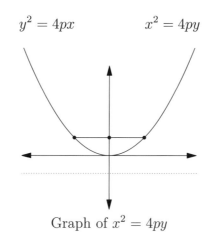

Graph of $x^2 = 4py$

As shown in the figure above, the line segment that passes through the focus, parallel to the **directrix**(the dotted line in the graph above telling us the direction of the curve), is known as the **latus rectum**, and half of this chord is known as **semilatus rectum**. Latus rectum has the length of $4|p|$ where $|p|$ is the distance between the vertex and the focus. Normally, we use semilatus rectum to graph a parabola.

14. Given $y = \dfrac{x^2}{2}$, which of the following is the measure of the latus rectum?

(A) $\dfrac{1}{2}$ 　　　　 (B) 1 　　　　 (C) 2 　　　　 (D) 4 　　　　 (E) 8

15. If $y = 2x^2 + 4x - 3$, which of the following is the directrix of the parabola?

(A) $y = -\dfrac{1}{8}$ 　　 (B) $y = -\dfrac{1}{4}$ 　　 (C) $y = -\dfrac{39}{8}$ 　　 (D) $y = -\dfrac{41}{8}$ 　　 (E) $y = -\dfrac{39}{4}$

Ellipse is the set of points whose sum of distances from two fixed points is invariant. Remember the given form of an ellipse is centered at $(0,0)$.

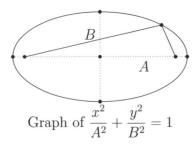

$$\text{Graph of } \frac{x^2}{A^2} + \frac{y^2}{B^2} = 1$$

We can use the rectangle drawn using A and B to fit an ellipse inside the rectangle. As the equation above suggests, the signs of leading coefficients of x^2 and y^2 are equal to one another. For instance, $3x^2 + 4y^2 = 5$ is an ellipse equation. Here are some properties related to ellipse.

- The length of major axis : Choose the larger of the two between A and B, and multiply it by 2. This number is important because this is the length that equals the sum of distances from two foci. In our figure above, the length must be $2A$, not $2B$.

- The length of minor axis : Choose the smaller of the two between A and B, and multiply it by 2. In the figure above, the length must be $2B$.

- The position of foci : shift $\pm c$ from the center either left/right or up/down, according to the given form, where $c = \sqrt{\max(A,B)^2 - \min(A,B)^2}$. Since the major axis is horizontal, foci must lie on the horizontal line. As one could see from the figure above, the sum of distances from two foci, denoted by two solid line segments, is always unchanging, and its measure equals the length of the major axis.

Let's have a look at the properties above, in detail, with the following example. For instance, suppose $\frac{x^2}{16} + \frac{y^2}{9} = 1$ is given to us. The length of major axis is $2 \times 4 (= 8)$. Similarly, the length of minor axis is $2 \times 3 (= 6)$. Lastly, the position of foci shifts $\sqrt{7}$ units left / right from the center $(0,0)$, so that two foci are located at $(-\sqrt{7}, 0)$ and $(\sqrt{7}, 0)$.

16. Find the length of major axis of $x^2 - 2x + 2y^2 + 4y = 5$.

Hyperbola is the set of points whose difference of distances from two fixed points is invariant. As in the figure below, the difference of upper segment lengths equals that of lower segment lengths, i.e., $d_2 - d_1 = d_3 - d_4$. Remember the given form of hyperbola is centered at the origin, which can be given by two different forms that start with either x^2 or y^2.

$$\frac{x^2}{A^2} - \frac{y^2}{B^2} = 1 \text{(left/right)} \qquad \frac{y^2}{B^2} - \frac{x^2}{A^2} = 1 \text{(up/down)}$$

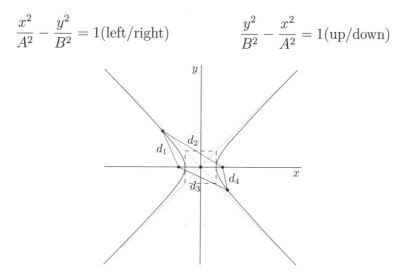

The figure above is the graph of $x^2 - y^2 = 1$. Check out how we graph a hyperbola.

1. Sketch a box that is 1 unit left/right of $(0,0)$ and 1 unit up/down of $(0,0)$.
2. Draw two lines passing through the opposite vertices, which are known as asymptotes.
3. The foci are located at $(-\sqrt{2}, 0)$ and $(\sqrt{2}, 0)$.

There are two properties of hyperbola, given by $\dfrac{(x-h)^2}{A^2} - \dfrac{(y-k)^2}{B^2} = 1$.

- The equations for asymptotes : $y = \pm\dfrac{B}{A}(x - h) + k$. It may be bit off-topic to directly ask about asymptotes, but the asymptotes may be provided to students to see what they geometrically mean.

- The center of hyperbola: In the equation, (h, k) can be directly found. If the equation is not given to us, the point of intersection between the asymptotes is the center of the hyperbola.

17. Find the asymptotes of $x^2 - 2x - 2y^2 + 4y = 2$.

Core Skill Practice 1

01.

Find the distance d between the points $(1, 2)$ and $(4, 5)$.

02.

Find the area of the triangle $A(2, 1)$, $B(2, 3)$ and $C(3, 1)$.

03.

Find the midpoint of a line segment connecting $P(2, 3)$ and $Q(4, 5)$.

04.

Find all points on the x-axis that are 7 units from the point $(3, 6)$.

05.

If two vertices of an equilateral triangle are $(0, 2)$ and $(0, 0)$, find all possible third vertices.

06.

Determine whether the following points are on the graph of $4x + 2y = 12$. Specify the position of the point relative to the line.

(a) $(4, -3)$ (b) $(2, -3)$ (c) $(3, 0)$

07.

Graph the equation $2x + y = 3$ in the following coordinate plane.

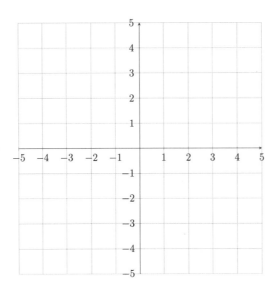

08.

Find the x-intercepts and y-intercept of the graph of $y = 8 - 2x^2$.

09.

Suppose a graph is symmetric with respect to the x-axis, and $(3, 5)$ is on the graph. Which of the following points must be on the graph?

(A) $(-3, 5)$
(B) $(3, -5)$
(C) $(-3, -5)$
(D) $(5, 3)$
(E) $(-5, 3)$

10.

Suppose a graph is symmetric with respect to the origin, and $(3, 5)$ is on the graph. Which of the following points must be on the graph?

(A) $(3, -5)$
(B) $(3, -5)$
(C) $(-3, -5)$
(D) $(-5, -3)$
(E) $(-5, 3)$

11.

Is $y = \dfrac{3x^2}{4 + |x|}$ symmetric with respect to x-axis, y-axis, or the origin?

(a) Check for the symmetry about the x-axis.

(b) Check for the symmetry about the y-axis.

(c) Check for the symmetry about the origin.

12.

Find the center of a circle $x^2 - 4x + y^2 - 6y = 10$.

13.

Find the asymptotes of a hyperbola $x^2 - 2x - y^2 + 4y = 2$.

14.

Compute the length of latus rectum of $y = 2x^2$.

15.

Find the length of major axis of $\dfrac{x^2}{6} + \dfrac{y^2}{7} = 1$.

Solution to Core Skill Practice 1

01.
$$d = \sqrt{(4-1)^2 + (5-2)^2}$$
$$= 3\sqrt{2}$$

02. Since $AB = 2$, and its height is given by 1, we get $1 \times 2 \times \dfrac{1}{2} = 1$.

03. Apply midpoint formula to get $M = \left(\dfrac{2+4}{2}, \dfrac{3+5}{2}\right) = (3,4)$.

04.
$$\sqrt{(x-3)^2 + (0-6)^2} = 7$$
$$(x-3)^2 + 36 = 49$$
$$(x-3)^2 = 13$$

Hence, we get points $(3 \pm \sqrt{13}, 0)$.

05. $(\pm\sqrt{3}, 1)$.

06.
(a) No. It is below the graph.

(b) No. It is below the graph.

(c) Yes. It is on the graph.

07.

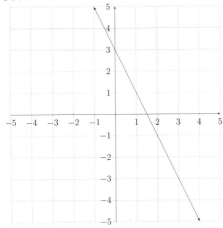

08. x-intercepts are $(\pm 2, 0)$; y-intercept is $(0, 8)$.

09. $(3, -5)$ must be on the graph. Therefore, the correct answer is (B).

10. $(-3, -5)$ must be on the graph. Thus, the answer must be (C).

11.

(a) Substitute $-y$ instead of y. The form is not equal to the original form, so the graph is NOT symmetric about the x-axis.

(b) The graph is symmetric about the y-axis.

(c) Substitute $(-x, -y)$ instead of (x, y). The form is not equal to the original form, so the graph is NOT symmetric about the origin.

12. Complete the square to get its center as $(2, 3)$.

13. Asymptotes are $y = \pm(x - 1) + 2$, so $y = x + 1$ or $y = -x + 3$.

14. Since $p = \dfrac{1}{8}$, the length of latus rectum is $\dfrac{1}{2}$.

15. Since $\sqrt{6} < \sqrt{7}$, the length of major axis is $2\sqrt{7}$.

TOPIC
2

Essence of Functions

Function is a term coined by Leibniz (1646-1716). In correspondence with Johann Bernoulli, who also described a function as a quantity formed from variable(or indeterminate) and constant quantities, Leibniz must have agreed with this idea. Later on, in 18th century, Euler, a student of Bernoulli, clarified it closer to what we know. Since then, we use a function as a relation in which none of the two different ordered pairs has the same x-coordinate or first component.

Mathematicians in early 1900's did not use the notion of **vertical line test**. They did not distinguish between which is function and which isn't. Although the exact source of where VLT began is unknown, the vertical line test did show up around 1930s. This test determines whether the given graph belongs to that of a function or relation. If we draw all possible vertical lines on the graph of a relation, the relation is

1. a function if each line cuts the graph no more than once.

2. not a function if at least one line cuts the graph more than once.

Notice that $y = f(x)$ or $f : x \longrightarrow y$ is a typical way of writing a function where

- *Domain*, a term of which was used by Arthur Cayley in late 1880's, is the set of x values (input).

- *Range*, a term of which was used by John Spare in late 1860's, is the set of actual y values (actual output).

1. Let $f(x) = 3x^2 - 2$.

(a) Find $f(2)$.

(b) Find all values of α such that $f(\alpha) = 47$.

(c) Find $f(2x - 1)$.

(d) Find $f(1 - 2x)$.

You can also put a function inside a function. In another section, we will see in detail of how a function can be placed inside another function, but we can algebraically compute how it looks like.

2. If $f(3x) = 4x^2$, which of the following equals $f(g(x))$ for some function $g(x)$?

(A) $\dfrac{4(g(x))}{3}$

(B) $\dfrac{4(g(x))^2}{3}$

(C) $\dfrac{4(g(x))^2}{9}$

(D) $\dfrac{16(g(x))^2}{3}$

(E) $\dfrac{16(g(x))^2}{9}$

Linear function is a surjective(or onto) function, meaning that if a bystander gives you a y-value, we can always find the corresponding x-value. This gives us a relieving sensation because we can always find one x-value for any y-value, especially when we are given with a linear function.

3. If $f(x) = 3x + 2$ and $g(x) = 2x - 3$, then which of the following equals 5?

(A) $f(g(1))$
(B) $g(f(1))$
(C) $f(g(2))$
(D) $g(f(2))$
(E) $f(g(0))$

Given two functions f and g, as though two numbers can be added, subtracted, multiplied, or divided, functions have the same four arithmetic well-defined. We will see how arithmetic plays an important role in graphing functions, especially in addition or subtraction.

- Addition

  $(f + g)(x)$ is defined by $f(x) + g(x)$, and the resulting expression should be simplified if possible. In terms of adding polynomials or some radical expressions, similar terms (or like terms) can be simplified. When sketching a graph of $f(x) + g(x)$, we must <u>add</u> the heights at a fixed value of x.

- Subtraction

  $(f - g)(x)$ is defined by $f(x) - g(x)$, and the resulting expression should be simplified if possible. In terms of subtracting polynomials or some radical expressions, similar terms (or like terms) can be simplified. When sketching a graph of $f(x) - g(x)$, we must <u>subtract</u> the heights at a fixed value of x.

- Multiplication

  $(f \times g)(x)$ is defined by $f(x) \times g(x)$, and the resulting expression should be simplified if possible. Techniques are FOIL, only in special product of binomials, or the application of general distributive law. When sketching a graph of $f(x) \times g(x)$, we must look at the x-intercepts of individual functions to figure out the new x-intercepts, and see how the graph may behave in between the x-intercepts, if any.

- Division

  $\left(\dfrac{f}{g}\right)(x)$ is defined by $\dfrac{f(x)}{g(x)}$ where $g(x) \neq 0$, and the resulting expression should be simplified if possible. Its graph must be that of rational functions, so we use the same strategies employed in graphing rational functions.

4. Let $f(x) = \dfrac{x+4}{x-4}$. If $|x| \neq 4$, graph $f(x) \cdot f(-x)$. Explain how it is different from the graph of a constant function.

5. If $f(x) = x^2 + 4x - 2$ and $g(x) = 2x^2 - x - 4$, then $(f + g)(x) =$

(A) $(x + 3)(3x - 2)$
(B) $(3x + 2)(x - 3)$
(C) $3(x - 1)(x + 2)$
(D) $3(x + 1)(x - 2)$
(E) $(x + 3)(3x + 2)$

6.

(a) Which of the following expressions is equivalent to $f(x) = \dfrac{x^2 + x + 1}{x - 1}$?

(A) $(x + 2) + \dfrac{3}{x - 1}$ (B) $(x - 1) + \dfrac{3}{x - 2}$ (C) $(3x - 1) + \dfrac{1}{x - 2}$ (D) $(3x + 1) - \dfrac{6}{x - 1}$

(b) For $x > 1$, find the minimum point of $f(x)$ using AM-GM inequality.[1]

---

[1]Given two positive real numbers a, b, AM-GM inequality states that $\dfrac{a + b}{2} \geq \sqrt{ab}$, where equality holds if $a = b$.

7. Sketch the graph of $f(x) = (x+2) \cdot |x-1|$, using the graph of $y = x+2$ and $y = |x-1|$, as shown in the figure below.

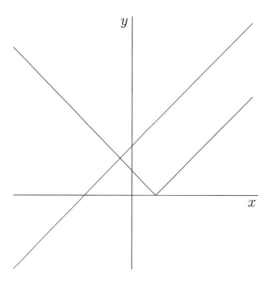

8. If $f(x) = \lfloor x \rfloor$, draw $g(x) = x^2 f(x)$ for $-2 \leq x \leq 2$.

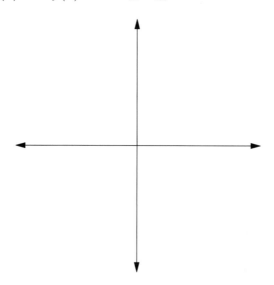

First, let's review what we learned in Algebra 2. We may start from linear function to exponential function.

Linear Function

$$f(x) = mx + n$$

where m is the slope and n is the y-intercept. Slope is the ratio between rise and run. Mathematically, it is the change of y-values with respect to the change of x-values.

Example

Find a linear function, given a slope of 4 and a point $(1, 3)$.

Solution

$$y = 4x + n$$
$$3 = 4(1) + n$$
$$-1 = n$$
$$y = 4x - 1$$

Let's analyze the domain and range of a linear function.

- Domain : $\mathbb{R}$

- Range : $\mathbb{R}$

If the range is the set of all real numbers $\mathbb{R}$, then we say the function is *surjective*, or *onto*. It means that any y in the codomain is assigned to some x in the domain. Not all functions are surjective. For an instance, think about $y = x^2$. If $y = -1$, no real x is assigned to $y = -1$. In other words, $y = x^2$ has its range a proper subset of its codomain $\mathbb{R}$.[2] Back to basic Algebra 2 level, let's refresh on the concept of *slope*. The slope is given by two points (x_1, y_1) and (x_2, y_2), which is defined by

$$m = \frac{\text{rise}}{\text{run}} = \frac{\triangle y}{\triangle x} = \frac{y_2 - y_1}{x_2 - x_1}$$

The most important concept we should remember from Precalculus, especially with respect to a linear function, is the point-slope form[3]. The equation of any line passing through the point (x_1, y_1) with the slope of m is given by $y = m(x - x_1) + y_1$. In Algebra 2, we are used to the slope-intercept form of $y = mx + b$, but onward, the slope-point form is more useful in both Precalculus and Calculus.

---

[2]The set of all *possible* y-values is the codomain of a function, but the set of *actual* y-values is the range of the function.

[3]There are two major forms for us to know : slope-intercept form and point-slope form.

(a) Given a linear function $f(x) = 3x + 4$, find the following functions.

(i) a linear function perpendicular to $f(x)$ that passes through $(0,0)$.
(ii) a linear function parallel to $f(x)$ that passes through $(1,4)$.

Solution
(i) $y = -\dfrac{1}{3}x$
(ii) $y = 3x + 1$

(b) Find the equation of the line in slope-intercept form of the following line with the slope information, passing through the given point.

$$m = 2, (4, -2)$$

Solution
$y - (-2) = 2(x - 4) \implies y = 2x - 8 - 2 = 2x - 10.$

(c) Find the equation of the line in slope-intercept form, passing through two points.

$$(2, 3) \text{ and } (4, 1)$$

Solution
$y - 3 = \dfrac{1-3}{4-2}(x - 2) \implies y = -(x - 2) + 3 = -x + 5.$

(d) Find the x-intercept and y-intercept of the line equation $2x + 3y - 12 = 0$.

Solution
$2x + 3y - 12 = 0 \implies \dfrac{x}{6} + \dfrac{y}{4} = 1 \implies (6, 0) \text{ and } (0, 4).$

9. Let p and q be integers. If the line $y = \dfrac{p}{\sqrt{2}+1}x + \dfrac{q}{\sqrt{2}-1}$ passes through $(\sqrt{2}-1, 10)$, find the sum of p and q.

(A) 2   (B) 3   (C) 4   (D) 5   (E) 6

10. Solve the following two part problems.

(a) If $y = mx + n$, where $m \neq 0$, if you give me a y-value, I can always find the corresponding x-value, due to (surjective / injective) property of a linear graph.

(b) If $y = f(x) = mx + n$, where $m \neq 0$, if $a \neq b$, then $f(a) \neq f(b)$, due to (surjective / injective) property of a linear graph.

As a recap, we can understand the meaning of surjective or injective function as in the following manner.

- Surjective function : the size of codomain is at most the size of domain. This implies that the size of all possible x-values might be equal to or greater than that of all possible y-values.

- Injective function : the size of domain is at most the size of codomain. This implies that the size of all possible y-values might be equal to or greater than that of all possible x-values.

11. If $f(x) = 3x + 2$ and $g(x) = -x + 3$, which of the following could be -2?

(A) $f(g(0))$
(B) $f(g(1))$
(C) $f(g(2))$
(D) $g(f(1))$
(E) $g(f(2))$

12. Let f be a linear function such that $f(5) - f(-5) = 2$. Which of the following must be $f(15) - f(-15)$?

(A) 2
(B) 3
(C) 4
(D) 6
(E) 8

01.
Find the general form of the line equation, containing $(4, -2)$ and parallel to the line containing $(2, 3)$ and $(-1, 4)$.

02.
Find k if $3x - 5y = 7$ and $2x - ky = 5$ are

(a) parallel.   (b) perpendicular.

03.
Find the line equation

(a) normal(or perpendicular) to the given line $y = 3$, passing through $(2, 1)$.

(b) parallel to the given line $x = 1$, passing through $(-3, 2)$.

04.
Given $y = 3x + b$ that passes through $(1, 5)$, what is the value of x if $y = 1$?

(A) $\dfrac{1}{3}$ (B) 2   (C) -2 (D) $-\dfrac{1}{3}$

Solution to Core Skill Practice 2

01. Slope can be found as $\dfrac{4-3}{-1-2} = \dfrac{1}{-3}$, so $y = -\dfrac{1}{3}(x-4) - 2$. Hence,
$-3y = x - 4 + 6 \implies x + 3y + 2 = 0$.

02.
(a) $\dfrac{2}{k} = \dfrac{3}{5} \implies 3k = 10 \implies k = \dfrac{10}{3}$

(b) $k = -\dfrac{6}{5}$

03.
(a) $x = 2$

(b) $x = -3$

04. The correct answer is (D). First, $5 = 3(1) + b \implies b = 2$. Since $y = 3x + 2$ is onto function, there exists x for $y = 1$. Hence, $1 = 3x + 2 \implies -1 = 3x \implies x = -\dfrac{1}{3}$.

Quadratic Function

$$f(x) = ax^2 + bx + c$$

is a quadratic function in standard form, where $a \neq 0$.

- $a > 0$ means that the graph of $y = ax^2 + bx + c$ is concave up.

- $a < 0$ means that the graph of $y = ax^2 + bx + c$ is concave down.

- c is the y-intercept : $f(0) = c$

There are two other forms of quadratic function for you to remember.

- Intercept form[4] : $y = a(x - p)(x - q)$ where p and q are real numbers. Here, $(p, 0)$ and $(q, 0)$ are the x-intercepts.

- Vertex form[5] : $y = a(x - h)^2 + k$ where h and k are real numbers. Here (h, k) is the coordinates of the vertex.

Unlike linear functions, quadratic functions are neither $1 - 1$ functions nor *onto*. Typically, they are $2 - 1$ function[6], and negative y values are not assigned by any $x \in \mathbb{R}$. Let's analyze the domain and range of a quadratic function.

- Domain : $\mathbb{R}$

- Range : It depends.

When we deal with range, we use *trivial inequality*, known as $x^2 \geq 0$ for all $x \in \mathbb{R}$. The following example illustrates how to use trivial inequality.

Example

Find the range of the function $f(x) = 3(x - 2)^2 + 4$.

Solution
Since $(x - 2)^2 \geq 0$ for any real number x, then $3(x - 2)^2 \geq 0$. Therefore, $f(x) = 3(x - 2)^2 + 4 \geq 4$. Hence, the range of the function $f(x)$ is equal to $[4, \infty)$.

Given $y = ax^2 + bx + c$, the discriminant $b^2 - 4ac$ will count the number of real solutions the quadratic has. In the following example, the vertex is lower than the x-axis and the graph is concave up so that the graph intersects the x-axis twice at different points.

---

[4]In order to find the intercept form, you should be able to factorize the given expression.
[5]In order to find the vertex form, you should be able to complete the square.
[6]It means that two values of x are assigned to one value of y.

How many x-intercepts does $f(x) = x^2 + 3x + 1$ have?

Solution
Let's use discriminant : $3^2 - 4(1)(1) = 9 - 4 = 5 > 0$, which implies that there are two distinct real x-intercepts.

13. Given $a \in \mathbb{R}$, if $x^2 + ax + 3 = 0$, where r and s are solutions, which of the following is the minimum value of $r^2 + s^2$?

(A) -3 (B) -4 (C) -5 (D) -6 (E) -7

The following example demonstrates the process of factorization and the application of quadratic formula, all of which you learned in Algebra 2. Given a quadratic equation, sometimes it is difficult to factorize; hence, it may be time-saving for you to use quadratic formula. If the discriminant is less than 0, then the equation has no real solution. Refresh your memories by reading the solution. Try them yourself.

$$ax^2 + bx + c = 0 \implies x = \frac{-b \pm \sqrt{b^2 - 4ac}}{2a}$$

Example

Solve for x.
(a) $x^2 - 14x + 49 = 0$.

Solution
$(x - 7)^2 = 0 \implies x = 7$

(b) $x^2 - 81 = 0$.

Solution
$(x - 9)(x + 9) = 0 \implies x = -9$ or 9

(c) $2x^2 - 9x - 18 = 0$.

Solution
$(2x + 3)(x - 6) = 0 \implies x = -\dfrac{3}{2}$ or 6

Switching $y = ax^2 + bx + c$ into $y = a(x - h)^2 + k$ illustrates that the graph of a quadratic function has the line of symmetry, which means that two different x-values result in the same y-value.

14. If $f(x) = a(x - h)^2 + k$ and $f(3) = f(9)$, which of the following must be the value of h?

(A) 3   (B) 6   (C) 9   (D) 12   (E) -12

In the late 16th century, *François Viéte*, a French mathematician and lawyer, was one of the first mathematicians who introduced the use of letters to represent variables. He significantly advanced Algebra by investigating the relationship between the roots and coefficients of a polynomial.

15. If $x^2 + 3x + 4 = 0$ has roots, r and s, which of the following could be the quadratic equation with roots $r - 1$ and $s - 1$?

(A) $x^2 - 5x + 8$
(B) $x^2 + 5x - 8$
(C) $x^2 + 5x + 8$
(D) $x^2 + 8x - 5$
(E) $x^2 + 8x + 5$

Core Skill Practice 3

01.
Solve for x if $x^3 - 3x^2 - 4x + 12 = 0$.

02.
Solve for x.

(a) $\dfrac{x}{3} + \dfrac{3}{x} = \dfrac{5}{2}$.

(b) $x - \dfrac{17}{4} = -\dfrac{1}{x}$.

03.
Solve for real number x if $x^4 - 8x^2 - 9 = 0$.

04.

If $f(x) = 5x^2 + 3x + b$, for which of the following values of b will the quadratic function have no real solution?

(A) -3 (B) -2 (C) -1 (D) 0   (E) 1

05.

If $y = x^2 + kx - k$, for what values of k will the quadratic have one real solution?

06.

Solve for x if $2x^4 - 15x^3 + 18x^2 = 0$.

07.

Find the domain of $y = \dfrac{1 - 2x}{6x^2 + 5x - 6}$.

Solution to Core Skill Practice 3

01. $x^2(x-3) - 4(x-3) = 0 \implies (x^2-4)(x-3) = 0 \implies x = \pm 2$ or 3.

02.

(a) $6x\left(\dfrac{x}{3} + \dfrac{3}{x} = \dfrac{5}{2}\right) \implies 2x^2 + 18 = 15x \implies (2x-3)(x-6) = 0 \implies x = \dfrac{3}{2}$ or 6.

(b) $4x\left(x - \dfrac{17}{4} = -\dfrac{1}{x}\right) \implies 4x^2 - 17x = -4 \implies 4x^2 - 17x + 4 = 0 \implies (4x-1)(x-4) =$
$0 \implies x = \dfrac{1}{4}$ or 4.

03. $(x^2-9)(x^2+1) = 0 \implies (x+3)(x-3)(x^2+1) = 0 \implies x = \pm 3$.

04. The correct answer is (E). $3^2 - 4(5b) < 0 \implies 9 - 20b < 0 \implies 9 < 20b \implies \dfrac{9}{20} < b$.

05. $k^2 - 4(-k) = 0 \implies k^2 + 4k = 0 \implies k(k+4) = 0 \implies k = 0$ or -4.

06. $x^2(2x^2 - 15x + 18) = 0 \implies x^2(2x-3)(x-6) = 0 \implies x = 0, \dfrac{3}{2}, 6$.

07. $6x^2 + 5x - 6 \neq 0 \implies (3x-2)(2x+3) \neq 0 \implies \mathbb{R}$ such that $x \neq \dfrac{2}{3}$ and $-\dfrac{3}{2}$.

Polynomial Function

First, polynomial is a sum of monomials, which consist of a variable, a number, or a product of variable and number.

$$p(x) = a_n x^n + a_{n-1} x^{n-1} + a_{n-2} x^{n-2} + \cdots + a_1 x + a_0$$

Degree, identifying a polynomial, is the highest exponent of polynomial terms. Let's look at the following example.

Example

Determine the degree and leading coefficient of $4x^3 + 4x + 2$.

> **Solution**
> The degree is the highest exponent, so it is 3 in this case. The leading coefficient is the coefficient of the highest power term, which means it is 4.

There are two forms for polynomial functions.

- Intercept form : $y = a(x - p_1)^{m_1}(x - p_2)^{m_2} \cdots (x - p_n)^{m_n}$[7]

- General form : $p(x) = a_n x^n + a_{n-1} x^{n-1} + a_{n-2} x^{n-2} + \cdots + a_1 x + a_0$[8]

When a higher-power polynomial function is given in the intercept form, the following analysis could be presented.

- Multiplicity : at x-intercept, the multiplicity determines whether the graph is tangent to the x-axis or it crosses the x-axis.

    Even multiplicity : the graph is tangent to the x-axis.

    Odd multiplicity : the graph crosses the x-axis

- Overall behavior : it is determined by both the leading coefficient and the degree of the function.

    Even degree : if the leading coefficient is positive, then the overall graph is concave up. If the leading coefficient is negative, then it is concave down.

    Odd degree : if the leading coefficient is positive, the function is increasing overall. Otherwise, the function is decreasing overall.

---

[7] General form is transformed into intercept form by factorization. The reason why intercept form is useful is because it tells us where roots are.

[8] General form is always retrieved whenever polynomial product is expanded.

Let's analyze the domain and range of a polynomial function.

- Domain : $\mathbb{R}$

- Range : It depends on the degree.

    Even degree : it has either maximum or minimum.

    Odd degree : $\mathbb{R}$.

16. If $f(x)$ is a cubic function with the leading coefficient of 1, and $f(-1) = -2$, $f(0) = 1$, and $f(1) = 3$, which of the following must be true?

(A) There exists a zero between $x = -1$ and $x = 0$.
(B) There exists a zero between $x = 0$ and $x = 1$.
(C) The function is increasing for $-1 < x < 0$.
(D) There is a minimum value for $f(x)$.
(E) The function is increasing for all real values of x.

17. If $y = \dfrac{1}{20}(x + 1)^2(x - 3)^3$, the graph is

(a) (tangent to / cutting through) the x-axis at $x = 3$.

(b) (tangent to / cutting through) the x-axis at $x = -1$.

(c) Hence, it is a(an) (increasing / decreasing / neither increasing nor decreasing) function.

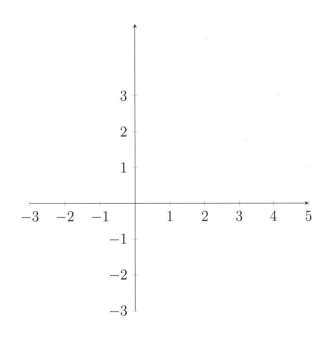

Factoring techniques are important upto Algebra 2 and Precalculus. The following forms are commonly used when we factor polynomials.

- $x^4 + x^3 + x^2 = x^2(x^2 + x + 1)$

- $x^2 - y^2 = (x - y)(x + y)$

- $x^2 + y^2 = (x + y)^2 - 2xy = (x - y)^2 + 2xy$

- $x^3 + y^3 = (x + y)(x^2 - xy + y^2) = (x + y)^3 - 3xy(x + y)$

Complete factorization usually refers to terms with rational coefficients. For example, $x^2 - 5$ can be factored into $(x - \sqrt{5})(x + \sqrt{5})$ with real number coefficients. However, since $\sqrt{5}$ is not rational, we simply acknowledge that $x^2 - 5$ cannot be factored into the product of two linear terms with rational coefficients, unless it is stated otherwise. Similarly, we say $x^2 + 1$ cannot be factored, although $(x + i)(x - i) = x^2 - i^2 = x^2 + 1$ with complex coefficients.

Example

Factor the following polynomial expressions.
(a) $8x^3 + 27y^3$.

Solution
$8x^3 + 27y^3 = (2x + 3y)(4x^2 - 6xy + 9y^2)$

(b) $x^4 - 11x^2 - 80$.

Solution
$x^4 - 11x^2 - 80 = (x^2 + 5)(x + 4)(x - 4)$

(c) $x^3 + 5x^2 - 4x - 20$.

Solution
$(x^3 + 5x^2) - 4(x + 5) = x^2(x + 5) - 4(x + 5) = (x^2 - 4)(x + 5) = (x + 2)(x - 2)(x - 5)$

(d) $16 - x^2 - 2xy - y^2$.

Solution
$16 - (x^2 + 2xy + y^2) = 4^2 - (x + y)^2 = (4 - x - y)(4 + x + y)$

(e) $x^4 + 3x^2 + 4$.

Solution
$x^4 + 4x^2 + 4 - x^2 = (x^2 + 2)^2 - x^2 = (x^2 - x + 2)(x^2 + x + 2)$

(f) $x^4 + 5x^2 + 9$.

Solution
$x^4 + 6x^2 + 9 - x^2 = (x^2 + 3)^2 - x^2 = (x^2 - x + 3)(x^2 + x + 3)$

An application of $a^3 + b^3 = (a+b)(a^2 - ab + b^2)$ and $(a+b)^3 - 3ab(a+b) = a^3 + b^3$ combines a cubic root expression to find out the value of some specific expressions.

- $a^3 + b^3 = (a+b)^3 - 3ab(a+b)$

- $a^3 - b^3 = (a-b)^3 + 3ab(a-b)$

We express the sum of cubes using $a+b$ and ab. Even when a question does not involve a and b, we simply label a given expression as a and b to use the above expressions.

18. If $\sqrt[3]{2 - x^3} + \sqrt[3]{2 + x^3} = 2$ for some positive real x, find the value of x^6.

Out of all other polynomial functions, a cubic function is slightly more important than others, since its property can be deduced easily from the graph. Have a look at the graph of $y = x^2(x - 3k)$.

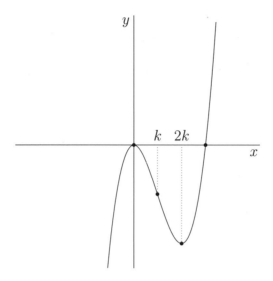

At $x = 0$, k, $2k$, and $3k$, we have the tangent point, the point of inflection (= the point of symmetry), the relative minimum, and the intersection point with the x-axis. You can guess why this is called cubic (degree 3) polynomial, since $[o, 3k]$ is cut into three congruent segments whose endpoints all refer to the important points in its graph.

19. Solve $27x^2(x - 1) + 4 = 0$ for $x < 0$, without using a calculator.

Core Skill Practice 4

Assume that the coefficients in this skill practice are rational numbers.

01.
Completely factor $x^3 - 25x$.

02.
Completely factor $3x^3 + 5x^2 + 2x$.

03.
Completely factor $3x^4 - 3$.

04.

Completely factor $16x^4 - 24x^2y + 9y^2$

05.

Completely factor $9a^4 - 36a^2b^2$.

06.

Completely factor $4x^4 + 7x^2 - 36$.

07.

Completely factor $250x^3 - 128$.

08.

Completely factor $x^5 + 17x^3 + 16x$.

09.

Completely factor $x^3 - xy^2 + x^2y - y^3$.

10.

Completely factor $x^6 - 9x^4 - 81x^2 + 729$.

11.

Completely factor $x^2 - 8xy + 16y^2 - 25$.

12.

Completely factor $x^5 + x^3 + x^2 + 1$.

13.
Completely factor $x^6 - 1$.

14.
Completely factor $x^6 + 1$.

15.
Completely factor $x^4 + x^2 + 1$.

Solution to Core Skill Practice 4

01. $x^3 - 25x = x(x^2 - 25) = x(x + 5)(x - 5)$

02. $3x^3 + 5x^2 + 2x = x(3x^2 + 5x + 2) = x(3x + 2)(x + 1)$

03. $3x^4 - 3 = 3(x^4 - 1) = 3(x^2 - 1)(x^2 + 1) = 3(x - 1)(x + 1)(x^2 + 1)$

04. $16x^4 - 24x^2y + 9y^2 = (4x^2 - 3y)^2$

05. $9a^4 - 36a^2b^2 = 9a^2(a^2 - 4b^2) = 9a^2(a + 2b)(a - 2b)$

06. $4x^4 + 7x^2 - 36 = (x^2 + 4)(4x^2 - 9) = (x^2 + 4)(2x - 3)(2x + 3)$

07. $2(125x^3 - 64) = 2(5x - 4)(25x^2 + 20x + 16)$

08. $x(x^4 + 17x^2 + 16) = x(x^2 + 16)(x^2 + 1)$

09. $x(x^2 - y^2) + y(x^2 - y^2) = (x + y)(x^2 - y^2) = (x + y)(x - y)(x + y) = (x + y)^2(x - y)$

10. $x^4(x^2 - 9) - 81(x^2 - 9) = (x^2 - 81)(x^2 - 9) = (x - 9)^2(x + 9)$

11. $(x^2 - 8xy + 16y^2) - 25 = (x - 4y)^2 - 5^2 = (x - 4y - 5)(x - 4y + 5)$

12. $x^3(x^2 + 1) + (x^2 + 1) = (x^3 + 1)(x^2 + 1) = (x + 1)(x^2 - x + 1)(x^2 + 1)$

13. $x^6 - 1 = (x^3 - 1)(x^3 + 1) = (x - 1)(x^2 - x + 1)(x + 1)(x^2 + x + 1)$

14. $x^6 + 1 = (x^2 + 1)(x^4 - x^2 + 1)$

15. $x^4 + x^2 + 1 = (x^2 + 1)^2 - x^2 = (x^2 - x + 1)(x^2 + x + 1)$

Given a polynomial function, how do we find integer roots? Finding integer roots can be extremely difficult and challenging work for us. The first example shows the first criteria of figuring out whether we have integer roots.

Example

Given $f(x) = x^4 + x^3 + x^2 - 15$, determine whether $f(2)$ is a root or not without evaluating $f(2)$.

Solution

If you look at $f(2) = 2^4 + 2^3 + 2^2 - 15$, the first three expressions are even, while the last term is odd. This means that $f(2) = 0$ is impossible to reach.

If we extend this idea, we can find rational roots, if any, for the given polynomial equation. Recall that a rational number is a quotient of integers. As we found integer roots, we could find rational roots, which will be handy when we are not allowed to use a calculator.

Rational Root Theorem

Let $f(x)$ be the polynomial

$$f(x) = a_n x^n + a_{n-1} x^{n-1} + a_{n-2} x^{n-2} + \cdots + a_1 x + a_0,$$

where all the a_i are integers, and both a_n and a_0 are nonzero. If p and q are relatively prime integers and $f(p/q) = 0$, then $p \mid a_0$ and $q \mid a_n$.

The method of finding rational roots is not specified in detail in many Precalculus (or Algebra 2) textbooks[9], but the following illustration will help you locate rational roots quickly.

Consider $p(x) = 2x^2 - 5x + 2$. We know that the roots are $x = 1/2$ and $x = 2$. Let's learn how to locate the roots using rational root theorem. What are the possible rational roots? According to the theorem, $x = \pm 1/2, \pm 1, \pm 2$. First, substitute $x = 0$ to find out that $p(0) = 2 > 0$. The graph is above the x-axis at $x = 0$. What is the next step for us? The next step is to determine whether roots are positive or negative. Since coefficients alternate in signs, roots must be positive. Suppose roots are negative, then simply putting $(-)$ in x tells us that $2(-)^2 - 5(-) + 2 = 2(+) + (+) + 2 = (+) > 0$. The assumption that roots are negative must be wrong.

Let's put $x = 1$ into $p(x)$, i.e., $p(1) = 2 - 5 + 2 = -1 < 0$. The graph is below the x-axis at $x = 1$. In other words, the graph must have passed the x-axis at least once between 0 and 1 because the sign of y-value changes as the graph moves from $x = 0$ to $x = 1$. The only possible rational root between 0 and 1 is $1/2$. Then,

$$p(1/2) = 2(1/4) - 5(1/2) + 2 = 1/2 - 5/2 + 2 = 5/2 - 5/2 = 0$$

---

[9]The only memorable book I found, which illustrates this process, is *Intermediate Algebra* by Richard Rusczyk and Matthew Crawford, the famous highschool math series by the Art of Problem Solving.

Voila! We located one root. Apply synthetic division to $p(x)$.

$$\begin{array}{r|rrr} 1/2 & 2 & -5 & 2 \\ & & 1 & -2 \\ \hline & 2 & -4 & 0 \end{array}$$

Hence, $p(x) = 2x^2 - 5x + 2 = (x - \frac{1}{2})(2x - 4) = 2(x - \frac{1}{2})(x - 2) = (2x - 1)(x - 2)$.

20.

(a) Solve $2x^3 - 6x^2 - 20x + 48 = 0$ by finding possible integer roots.

(b) Solve $12x^3 - 107x^2 - 15x + 54 = 0$ by finding all possible rational roots.

Synthetic division by *Paolo Ruffini* in early 19th century was invented to find out a root of polynomials. However, there is another property related to synthetic division. Applying synthetic divisions multiple times helps us find out the horizontal shift of a given polynomial function. We can perform horizontal shift by *direct substitution* or *synthetic division*.

21. Find $p(x)$ where $f(x + 1) = p(x)$ and $f(x) = x^3 + 3x^2 + 3x + 1$.

Also, upperbounds or lowerbounds are helpful for locating the roots of polynomials. Synthetic division can be used to determine upperbounds or lowerbounds on the roots of the given polynomials. Let's look at the following example.

Example

If $f(x) = x^3 + 10x^2 + 15x + 30$, how do we know there is no positive solution for which $f(x) = 0$?

> **Solution**
> Since all coefficients are positive, if we put any positive x into $f(x)$, it cannot be 0.

22.

(a) Find the quotient and remainder when $f(x) = x^3 - 2x^2 + 4x - 3$ is divided by $x - 3$.

(b) Explain how the quotient and remainder tell us that there is no real root of $f(x)$ greater than 3.

All polynomial functions have their graphs connected. This means that we can plot a function without taking a pencil off the plane. An application of this property is called the *Intermediate Value Theorem*. This so-called IVT guarantees the existence of a x-value, but it does not locate where such x-value is. Oftentimes, math teachers call this *location principle*, since plugging consecutive integers and finding out the results having opposite signs may locate the real root(s).

23. Given $p(x) = 2x^3 - 3x^2 - 11x + 6$, describe how there is a root between 0 and 1, and find the root between 0 and 1 using rational root theorem.

Core Skill Practice 5

01.

Solve $2x^3 - x^2 - 2x + 1 = 0$ by finding rational roots.

02.

Solve $x^3 - 14x^2 - 16x + 15 = 0$ by finding rational roots.

03.

Solve $x^3 - 3x^2 - 10x + 24 = 0$ by finding rational roots.

04.

Solve $6x^3 - 5x^2 - 22x + 24 = 0$ by finding rational roots.

05.

Find all roots of $p(x) = 12x^3 - 28x^2 - 9x + 10$.

06.

Find all roots of $p(x) = 45x^3 - 18x^2 - 5x + 2$.

Solution to Core Skill Practice 5

01. $(2x - 1)(x - 1)(x + 1) = 0 \implies x = 1/2, \ 1, \ -1.$

02. $(x - 15)(x^2 + x - 1) = 0 \implies x = 15, \ \dfrac{-1 \pm \sqrt{5}}{2}.$

03. $(x - 2)(x + 3)(x - 4) = 0 \implies x = 2, \ -3, \ 4.$

04. $(x + 2)(3x - 4)(2x - 3) = 0 \implies x = -2, \ \dfrac{4}{3}, \ \dfrac{3}{2}.$

05. $(2x - 1)(2x - 5)(3x + 2) = 0 \implies x = \dfrac{1}{2}, \ \dfrac{5}{2}, \ -\dfrac{2}{3}.$

06. $3\left(x - \dfrac{1}{3}\right)(3x + 1)(5x - 2) \implies x = \dfrac{1}{3}, \ -\dfrac{1}{3}, \ \dfrac{2}{5}.$

Radical Function

Radical functions can be categorized into two functions, even-indexed and odd-indexed. The family of even-indexed radical functions are similar to one another while different from the family of odd-indexed functions. Odd-indexed radicals are of form $\sqrt[3]{x}, \sqrt[5]{x}, \cdots$, while even-indexed radicals are of form $\sqrt{x}, \sqrt[4]{x}, \sqrt[6]{x}, \cdots$. Yet, when a real x is positive, $\sqrt{x}$ or $\sqrt[3]{x}$ are both concave down for their graphs.

- Even-indexed Function : if a radical function is even-indexed, then the domain depends on the epxression inside the radicand. Usually, the function $f(x) = \sqrt{x}$ is graphed only in the first quadrant.

- Odd-indexed Function : if a radical function is odd-indexed, then the domain is the set of all real numbers. The function $f(x) = \sqrt[3]{x}$ is graphed overall, most of which is drawn at the first or third quadrant.

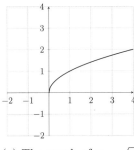

(a) The graph of $y = \sqrt{x}$

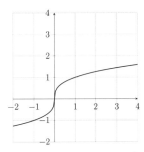

(b) The graph of $y = x^{\frac{1}{3}}$

Example

Find the domain and range of $f(x) = \sqrt{3-x} + 5$.

Solution
First, the domain depends on the radicand, i.e., $3-x \geq 0$. Hence, $3 \geq x$ is the condition for the domain, which means that $(-\infty, 3]$ is the domain. On the other hand, the range depends on the radical expression.

$$\sqrt{3-x} \geq 0$$
$$\sqrt{3-x} + 5 \geq 5$$

As one can see from the graph, a root function $y = \sqrt{x}$ has two different behaviors shown in the domain. If $0 \leq x < 1$, then the graph is above the line $y = x$. On the other hand, if $1 \leq x$, then the graph is lower than the line $y = x$. This means that if positive x is small enough, $y = \sqrt{x}$ spikes up faster than $y = x$, but goes slower than $y = x$ if x is large.

Another typical application of radical function is a question about the domain, especially when a function is composed with another function. Radical functions in GPA-related materials will have at most two functions with respect to composition, but radical functions in math competition will have multiple compositions.

If there is an expression about $\sqrt{(x-a)^2 + (y-b)^2}$, we may usually think of the distance between (x, y) and (a, b). Although finding algebraic solutions looks bit easier in radical function, geometric method may quickly present us with a more intuitive solution.

24. The minimum value of $\sqrt{(x-5)^2 + 9} + \sqrt{x^2 + 4}$ occurs at $x = p$. Find the minimum value.

Given multiple radical functions inside one another, we may be asked to find x-values or domain. We will cover the method of finding domain in latter part of this chapter, so let's have a look at the former one.

25. If $\sqrt{x + \sqrt{4x + \sqrt{16x + 1025}}} = 1 + \sqrt{x}$, then $x = p^q$, where p is a prime. Find $p + q$.

Absolute-value Function

Absolute-value function is a piecewise function that has two sub-domains. Given $f(x) = |x|$, if $x > 0$, then $f(x) = x$. Otherwise, $f(x) = -x$. The graph has a vertex at $(0,0)$, the lowest point of the graph in this case.

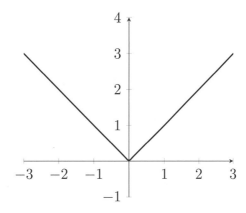

The graph of $y = |x|$

Example

Find the domain and range of the function $f(x) = 3|x - 2| + 3$.

Solution
Domain is $\mathbb{R}$. However, the range is not $\mathbb{R}$.

$$|x - 2| \geq 0$$
$$3|x - 2| \geq 0$$
$$3|x - 2| + 3 \geq 3$$

Two piecewise functions - absolute-value function and floor function - usually call for caseworks. For $y = |x|$, we study two possible cases : $x \geq 0$ or $x < 0$.

26. Find the sum of all possible solutions to $|x - 2| = |x - 3| + |x - 4|$.

27. Find the area of the region enclosed by $|3x| + |5y| = 15$.

28. Find the range of k-values where $f(x) = |x^2 + 4x - 3| = k$ has four intersection points, and graph the function $y = f(x) = |x^2 + 4x - 3|$.

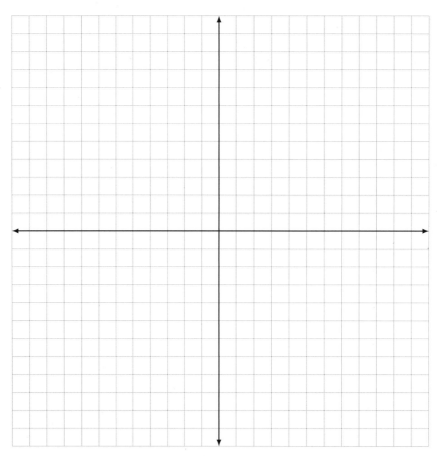

Typical conjunctive inequality may be combined with number theory, especially integer problems. Conjunctive inequality has the form of $|ax + b| < c$, and its solving method uses $-c < ax + b < c$. In particular, one may have to solve problems with the number of integer solutions inside a conjunctive inequality. Have a look at the following examples of analyzing intervals of size length 2, i.e., $(x, x + 2)$ for some real x.

- Let's say we have $(0.1, 0.1 + 2)$. Then, $0.1 < n < 2.1$ has two integer values of n satisfying the given inequality.

- Let's say we have $(0, 2)$. Then, $0 < n < 2$ has one integer value of n satisfying the given inequality.

- Let's say we have $(\frac{\sqrt{2}}{2}, 2 + \frac{\sqrt{2}}{2})$. Then, $0 + \frac{\sqrt{2}}{2} < n < 2 + \frac{\sqrt{2}}{2}$ has two integer values of n satisfying the given inequality.

29. For integers $1 \leq n \leq 2021$, the inequality $|\frac{nx}{2021} - 1| < \frac{n}{2021}$ has one integer solution x. How many n's are there? (One may notice that $\frac{nx}{2021} - 1$ is an integer.)

30. Find the integer solution to $||x - 10| - 2| = ||x - 8| - 4|$ if $x < 8$.

Floor function

Floor function is also known as the step function. The function value of $f(x) = \lfloor x \rfloor$ is the greatest integer less than or equal to x. Let's follow the following logic to find $\lfloor x \rfloor$ easily.

1. Plot the point x in the real line.

2. Consider all integers left of x.

3. Find the closest of all.

For example, in order to find $\lfloor -5.23 \rfloor$, list all those numbers that are less than -5.23. It is $-6, -7, -8, \cdots$. Out of all these numbers, -6 is the greatest. Therefore, $\lfloor -5.23 \rfloor = -6$.

The graph of the floor function is not continuous, which means that the graph is all chopped off. Let's analyze the domain and the range of the function.

- Domain : $\mathbb{R}$

- Range : $\mathbb{Z}$, the set of all integers.

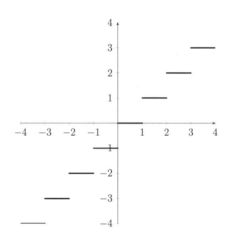

The graph of $y = \lfloor x \rfloor$

Example

Find the range of $y = \lfloor x \rfloor + 1$.

Solution
Since the graph of $y = \lfloor x \rfloor$ is shifted up by 1 unit, the range does not change. Therefore, it is the set of all integers, i.e., $\mathbb{Z}$.

31. Graph the following function such that $f(x) = 1$ for $1 \leq x < 2$ and $f(x-1) + 1 = f(x)$ for all real x. Compare the graph with the graph of $y = \lfloor x \rfloor$.

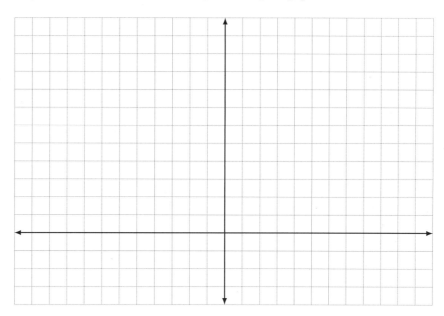

32. Graph the following function such that $f(x) = 2$ for $-2 \leq x < 1$ and $2f(x-3) = f(x)$ for all real x. Graph the function for $-5 \leq x < 7$.

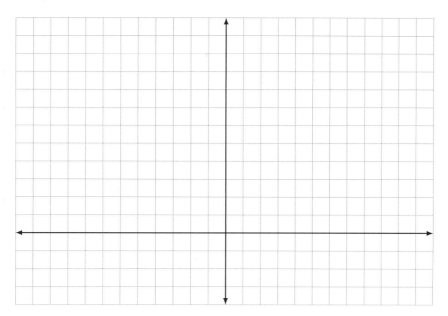

Rational Function

Given a rational function $y = \dfrac{p(x)}{q(x)}$, one must find undefined values of x by looking at the denominator. There are rules to remember before finding these undefined values. If $y = \dfrac{p(x)}{q(x)}$ has an equivalent, reduced form $y = \dfrac{p'(x)}{q'(x)}$, then the following criteria is helpful.

1. Before factorization(or reduction), find all values of x such that $q(x) = 0$.

2. After factorization, find values of x such that $q'(x) = 0$.

3. Whatever that is left behind is the vertical asymptote; otherwise, it is a hole.

Example

Given $f(x) = \dfrac{x+2}{x^2 + x - 2}$, find the domain of the function $y = f(x)$.

Solution
First, $y = \dfrac{x+2}{(x+2)(x-1)}$ tells us that the domain is all real numbers except $x = -2$ and $x = 1$. Specifically, $x = 1$ is a vertical asymptote, whereas $x = -2$ is a hole.

The range is more intricate. It uses the trick of substitution, using both quadratic function and linear function.

Example

Find the range of $y = \dfrac{2x}{x+3}$.

Solution
Let $\dfrac{2x}{x+3} = k$. Then $2x = kx + 3k$. Hence, $2x - kx = 3k$. So, $x = \dfrac{3k}{2-k}$. Since x should be a real number, the denominator should not be equal to 0. Hence, $k \neq 2$. Therefore, the range is the set of all real numbers except $y = 2$.

If a rational function is a fraction of linear forms, then it is not that difficult. In Algebra 2, when rational functions are covered, we learn that horizontal asymptotes are usually the y-values excluded from the range. The following skill practice reviews on the method of finding domain and range of rational functions in linear terms.

How about the range of rational function with quadratic terms? We use discriminant to find the range. Given $ax^2 + bx + c = 0$, there are three possible cases of discriminant analyses.

- Discriminant > 0 : there are two distinct real solutions.

- Discriminant $= 0$: there is one real solution.

- Discriminant < 0 : there is no real solution.

We will use the fact that the rational function has a real number input. In other words, all possible x values in the rational function are real numbers, i.e., discriminant must be at least 0. Let's see how this is used when we solve the following exercise.

33. Find the range of the function $y = f(x) = \dfrac{2x^2 + 3x + 1}{x - 1}$.

34. Find the number of integers in the domain of $f(x)$ where

$$f(x) = \frac{\sqrt{\sqrt{80 - \sqrt{x - 5000}} - 3}}{(x - 5000)(x - 5001)(x - 5002) \cdots (x - 5040)}$$

What if the rational function sits inside another rational function? How do we find the domain or range? First, we should look at the domain. Then, we find y-values for corresponding x-values.

35. Find the domain of $g(x) = \dfrac{3x - 2}{\dfrac{2}{x} - \dfrac{3}{3x - 4}}$

As a recap, let's review what we covered so far.

- Linear function is bijective, both onto and injective.

- Quadratic function's range depends on the leading coefficient. If the leading coefficient is positive, then the range has its minimum; if negative, then the range has its maximum.

- Polynomial function's range depends on the degree. If degree is even, it has either maximum or minimum. If degree is odd, the function is onto.

- Radical function's domain and range depend on the index. If index is even, then they are both restricted. Otherwise, they are both $\mathbb{R}$.

- Absolute valued function's range is restricted, i.e., $|x| \geq 0$.

- Floor function's range is $\mathbb{Z}$.

- Rational function's range varies. In order for real y to exist, the corresponding real x must exist.

Restriction of domain changes range as well. The purpose of restriction is to make a function $y = f(x)$ into a 1-to-1 function so that we could find an inverse function $y = f^{-1}(x)$. When domain is restricted, make sure which functions are increasing or decreasing. Typical examples of 1-to-1 functions are linear function, cubic function, radical function, exponential function and logarithmic function. Have a look at the following examples to see how restricted domain affects the range for logarithmic and exponential functions.

Example

Part 1. If $4 \leq x \leq 8$, find the range of $y = f(x) = \log_2(x) + 3$.

Solution
If $4 \leq x \leq 8$, then $\log_2 4 \leq \log_2(x) \leq \log_2 8$, so $2 \leq \log_2(x) \leq 3$. Hence, $5 \leq f(x) \leq 6$, i.e., $[5, 6]$.

Part 2. If $3 \leq x \leq 5$, find the range of $y = g(x) = (1/2)^x + 1$.

Solution
If $3 \leq x \leq 5$, then $(1/2)^3 \geq (1/2)^x \geq (1/2)^5$ because $(1/2)^x$ is monotonically decreasing. Hence, $1/32 \leq (1/2)^x \leq 1/8$. Therefore, $33/32 \leq g(x) \leq 9/8$, i.e., $[33/32, 9/8]$.

36. Suppose $p(x) = 3\sqrt{x} - 2$, but that $p(x)$ is only defined for $4 \leq x \leq 9$. In other words, the domain of p is the set of all real numbers from 4 to 9.

(a) Is there a value of x for which $p(x) = 3$?　　　　(b) What is the range of $p(x)$?

01.

Let $f(x) = x^2 - 2x - 8$.

(a) Compute $f(4)$.

(b) Find all x such that $f(x) = 0$.

(c) Find $f(x+1)$.

(d) Is there a real value of x such that $f(x) = -10$?

(e) Find the range of $f(x)$.

02.

Find the domain and range of each of the following real-valued functions.

(a) $f(x) = |1 - x|$

(b) $f(x) = \sqrt{2 - x}$

(c) $f(x) = 2 - x^2$

(d) $f(x) = \dfrac{1}{1 + \dfrac{1}{x}}$

03.

Let $f(x) = \sqrt{5-x}$ and $g(x) = \sqrt{2x-8}$. Find the domain of $f \cdot g$ and f/g, respectively. Also, find the domain of the product of two functions if $f(x) = \sqrt{3-x}$ and $g(x) = \sqrt{2x-8}$.

04.

Let $f(x) = \sqrt{\dfrac{x-2}{x-4}}$ and $g(x) = \dfrac{\sqrt{x-2}}{\sqrt{x-4}}$. Find the domain of $f(x)$ and $g(x)$, respectively.

05.

A function $h(x)$ is defined for $[-2, 4]$ such that $h(x) = (x-1)^2$. Find the range of $h(x)$.

06.

Find the quadratic function that has the vertex $(1, 4)$, passing through a point $(-2, 1)$.

07.

Write the quadratic function in vertex form $y = 2x^2 + 3x + 5$.

08.

Find a polynomial function with integer coefficients and the leading coefficient of 1, with the zeros $x = 3$ and $x = 1 \pm i$.

09.

Find the domain and range of the following rational functions.

(a) $f(x) = \dfrac{1}{\sqrt{x-5}}$

(b) $g(x) = \dfrac{2x^2 - x + 3}{x - 3}$

10.

Recall that any logarithmic function, which will be covered in the next topic, always has positive inputs. Hence, find the domain and range of the following logarithmic function

$$y = \log(x^2 - 4x + 5)$$

Solution to Core Skill Practice 6

01.
(a) $f(4) = 0$.
(b) $x = 4, -2$.
(c) $x^2 - 9$.
(d) No.
(e) $[-9, \infty)$

02.
(a) Domain : $\mathbb{R}$, Range : $[0, \infty)$.
(b) Domain : $(-\infty, 2]$, Range : $[0, \infty)$
(c) Domain : $\mathbb{R}$, Range : $(-\infty, 2]$.
(d) Domain : $\mathbb{R}$ except $x = 0, -1$, Range : $\mathbb{R}$ except $y = 0, 1$.

03. Domain of $f \cdot g$ is $[4, 5]$. On the other hand, the domain of f/g is $(4, 5]$. Lastly, the domain of the third function is an empty set. In other words, there is no real value of x that can go inside the $f(x) \cdot g(x) = \sqrt{3 - x} \cdot \sqrt{2x - 8}$.

04. Domain of $f(x)$ is $(-\infty, -2] \cup (4, \infty)$. On the other hand, the domain of $g(x)$ is $(4, \infty)$.

05. The range of $h(x)$ is $[0, 9]$.

06. $y = -\dfrac{1}{3}(x - 1)^2 + 4$.

07. $y = 2x^2 + 3x + 5 = 2(x + \dfrac{3}{4})^2 + \dfrac{31}{8}$.

08. $f(x) = (x - 3)(x - (1 + i))(x - (1 - i)) = (x - 3)(x^2 - 2x + 2) = x^3 - 5x^2 + 8x - 6$.

09.

(a) Domain is $\{x \in \mathbb{R} : x > 5\}$, and range is $\{y \in \mathbb{R} : y > 0\}$.

(b) Domain is $\{x \in \mathbb{R} : x \neq 3\}$, and range is $\{y \in \mathbb{R} : y \leq -1 \text{ or } 23 \leq y\}$.

10. Domain is $\mathbb{R}$, and range is $\{y \in \mathbb{R} : y \geq 0\}$.

Sometimes, we want to hook two (or more) machines together, taking the output from one machine and putting it into another. When we connect functions together like this, we are performing a composition of the functions. For example, the expression $f(g(x))$ means we put our input x into function g, and then take the output, $g(x)$, and put that into function f. A composition of functions may also be indicated with the symbol $\circ$. For example, when we write $h = f \circ g$, we define the function h such that $h(x) = f(g(x))$.

37. Suppose $f(x) = 2x + 5$ and $g(x) = 3\sqrt{x - 3}$.

(a) Find $f(g(3))$ and $g(f(3))$.

(b) Find a if $f(g(a)) = 17$.

(c) Compute $g(f(-5))$.

(d) Suppose that $h = g \circ f$. What is the domain of h?

38. Let $f(x) = \sqrt{x}$ and $g(x) = x^2$. Is it true that $f(g(x))$ and $g(f(x))$ are the same function? In order to answer this question, answer the following parts of the question.

(a) Find the domain of $f(g(x))$.

(b) Find the domain of $g(f(x))$.

39. If $f(x)$ has the domain $(-2, 2)$, then find the domain of $f\left(\dfrac{x+2}{x-2}\right)$.

40. If $f(x) = \{x\}$, where $\{x\} = x - \lfloor x \rfloor$, sketch the graph of $f(xf(x))$ for $x \geq 0$.

41. Let $f(x) = \sqrt{x}$. Find the number of real solutions to $f(x) - \lfloor f(x) \rfloor = \dfrac{x}{36}$.

Inverse Function

If $y = f(x)$ has an inverse function, the inverse function is denoted $f^{-1}(x)$. The graph of $y = f(x)$ and that of $y = f^{-1}(x)$ are symmetric with respect to $y = x$. Also, the composition of two inverse functions results in the identity function.

$$(f \circ f^{-1})(x) = x \qquad (f^{-1} \circ f)(x) = x$$

In order to have an inverse function, the original function $y = f(x)$ must be a $1 - 1$ function.

Example

Determine whether the following function is one-to-one.
(a) $f(x) = x^2$ (b) $f(x) = x^5$

Solution
(a) This is not $1 - 1$ function. In fact, this is $2 - 1$ function. Two values of x correspond to one value of y. There is no inverse.
(b) This is $1 - 1$ function. Hence, there is the inverse of $f(x)$.

42. If $f(x) = 2x - 3$ and $g(x) = \dfrac{1}{2}(x + 3)$, find the x-coordinate where $f(x) = g(x)$.

43. If $f(x) = mx + b$, and $f(x) = f^{-1}(x)$ for all real x, find the product of all possible values of m.

Finding the intersection between the curve $y = f(x)$ and $y = f^{-1}(x)$ is equal to finding that between $y = f(x)$ and $y = x$ since the graphs of inverse functions are symmetric about $y = x$. Sometimes, this idea of solving $f(x) = f^{-1}(x)$ by $f(x) = x$ may be used a bit more intricately in combination with reflection.

44. Let $P = (10, 6)$. Let L_1 be the locus of points equidistant from P and from the y-axis and L_2 be the locus of points equidistant from P and from the x-axis. Find the point of intersection whose x, y coordinates are less than 10 and 6, respectively.

Core Skill Practice 7

01.

Describe all constants a and b such that $f(x) = \dfrac{2x + a}{bx - 2}$ and $f(x) = f^{-1}(x)$ for all x in the domain of f.

02.

For which constants $a, b, c,$ and d does the function $f(x) = \dfrac{cx + d}{ax + b}$ have an inverse?

03.

Find all possible values of m and n such that $f(x) = mx + n$ satisfies $f(f(x)) = x$ for all x.

04.

If $f(f(x)) = x$ for all x for which $f(x)$ is defined and $f(0) = 2020$, find all roots of the equation $f(x) = 0$.

05.

Find the number of intersection points between the graph of $y = x - \lfloor x \rfloor$ and that of $y = |\frac{1}{3}x|$.

06.

Let $f(x) = \dfrac{2-x}{x+1}$ and $g(x) = \dfrac{3}{x+1}$, then evaluate $g(f(g(f(g(f(g(f(5)))))))))$.

🔆 Solution to Core Skill Practice 7

01. All real a and b such that $ab \neq -4$.

02. All real a, b, c, and d such that $ad - bc \neq 0$.

03. $(m, n) = (1, 0)$ or $(-1, n)$ for any real number n.

04. By symmetry, $x = 2020$ works. This is the only solution because self-inverse function is $1 - 1$ function.

05. There are five points of intersection between the two graphs.

06. Since $g(f(x)) = x + 1$, and there are four $g(f(x))$, we get 9.

Can you compare $1000\sqrt{1001}$ and $1001\sqrt{1000}$?

SOLUTION

#1. First, it is easy to notice that $\dfrac{1000}{1001}$ and $\dfrac{\sqrt{1000}}{\sqrt{1001}}$ are both less than 1.

#2. Let's recall how $y = \sqrt{x}$ looked like in $0 < x < 1$. It is concave down.

#3. This means that $\sqrt{\dfrac{1000}{1001}}$ is greater than $\dfrac{1000}{1001}$.

#4. We finally conclude that $1000\sqrt{1001} < 1001\sqrt{1000}$.

#1. $1000000\sqrt[6]{1000001} < 1000001\sqrt[6]{1000000}$.

#2. $2000\sqrt[3]{3000} < 3000\sqrt[3]{2000}$.

#3. $N\sqrt{N+1} < (N+1)\sqrt{N}$ for large integer value of N.

TOPIC
3

Revisiting Logs and Exponentials

Given $y = a^x$, there are two possible types of functions depending on the value of a.

- $a > 1 : y = a^x$ is increasing.

- $1 > a > 0 : y = a^x$ is decreasing.

Let's analyze the domain and range of the function.

- Domain : $\mathbb{R}$

- Range : It depends, starting from $a^x > 0$

Example

Find an exponential function that passes through $(1, 3)$ and $(2, 9)$.

Solution
As x is increased by 1 unit, y value is multipled by 3. Hence, the base must be 3. So, write down $y = a \cdot 3^x + b$. Then, $3 = 3a + b$ and $9 = 9a + b$. Hence, $6a = 6$, so $a = 1$ and $b = 0$. Therefore, the function is $y = 3^x$.

As one can check from the example above, exponential functions have a peculiar property. Given $y = a \cdot b^x + c$,

$$(x, y) \implies (x + 1, y \times b)$$

In short, the x-coordinate follows an arithmetic sequence, while the y-coordinate follows a geometric sequence. On the other hand, logarithmic function shows the opposite property, since logarithmic functions are inverse functions of exponential functions. Given $y = \log_a x$, there are two possible functions, just like an exponential function.

$\checkmark \, 1 < a : y = \log_a(x)$ is increasing $\qquad \checkmark \, 0 < a < 1 : y = \log_a(x)$ is decreasing

Let's analyze the domain and range of the function.

- Domain : $x > 0$ for $y = \log_a(x)$

- Range : $\mathbb{R}$

Example

Find a logarithmic function that passes through $(4, 5)$ and $(8, 6)$.

Solution
As x is multiplied by 2, y is increased by 1. Therefore, the base must be 2. Hence, $y = a \log_2(x) + b$. So, $5 = 2a + b$ and $6 = 3a + b$. Hence, $a = 1$ and $b = 3$. Thus, $y = \log_2(x) + 3$.

This is exactly the opposite of exponential function. Given $y = a \log_b(x) + c$,

$$(x, y) \Longrightarrow (x \times b, y + 1)$$

In short, the x-coordinate follows a geometric sequence, while the y-coordinate follows an arithmetic sequence.

The exponential function $y = b^x$ is either monotonically increasing or decreasing function, which means it is $1 - 1$ function. If y values are different, then x values are different.

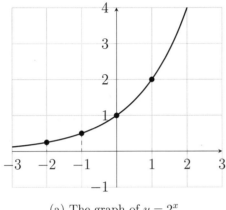

(a) The graph of $y = 2^x$

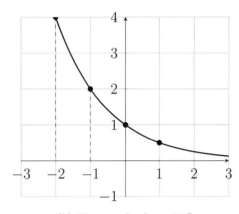

(b) The graph of $y = 2^{-x}$

On the other hand, logarithmic functions can be classified into two types. As y changes by 1 unit, the width changes by the amount shown in the base of the logarithm, as shown in the following figures.

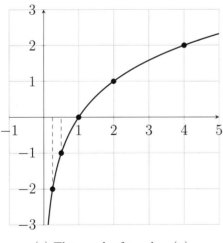

(a) The graph of $y = \log_2(x)$

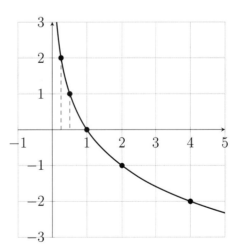

(b) The graph of $y = \log_{\frac{1}{2}}(x)$

1. Find the domain and range of the following functions.

(a) $f(x) = \dfrac{1 - 2^{x+1}}{2^x}$

(b) $f(x) = \log_2(x^2 - 4x + 3)$

2. Find the domain of $f(x) = \log_{\frac{1}{3}}(\log_9(\log_{\frac{1}{9}}(x)))$.

3. Sketch the graph of the following logarithmic functions.

(a) $y = \log_2 |x - 1|$

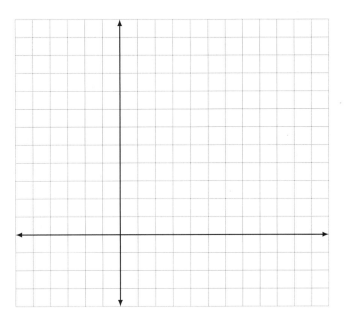

(b) $y = \log_2(x^2 - 6x + 10)$

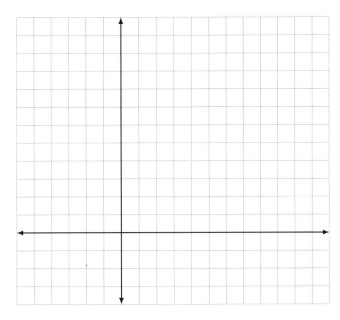

4. If $f(x) = x \cdot e^x$ for $x \geq 0$, it has an inverse function $f^{-1}(x)$. If $k \geq 0$ satisfies $ke^{2k} = e$, then $k = m \cdot f^{-1}(n)$. Find mn.

5. Find the number of solutions to $2^{x-1} = \lfloor \log_2 (x+1) \rfloor$ defined for $0 \leq x \leq 1$.

Let's refresh on what we know about exponential and logarithmic properties. Most of the properties are well-studied in Algebra 2. In this section, we will cover some challenging questions using exponential properties and logarithmic properties, suited for Precalculus level.

Exponential Properties

Given $a > 0$ where $a \neq 1$, the following properties work for rational m and n.

- $a^m a^n = a^{m+n}$

- $\dfrac{a^m}{a^n} = a^{m-n}$

- $(a^m)^n = a^{mn}$

- $a^0 = 1$

- $a^1 = a$

- $a^{-1} = \dfrac{1}{a}$

Logarithmic Properties

Given $a > 0$ where $a \neq 1$, the following properties work for positive real x and y.

- $\log_a(x) + \log_a(y) = \log_a(xy)$

- $\log_a(x) - \log_a(y) = \log_a\left(\dfrac{x}{y}\right)$

- $\log_a(x^y) = y \log_a(x)$

- $\log_{a^n}(b^m) = \dfrac{m}{n} \log_a(b)$

- $\log_a b = \dfrac{\log_c b}{\log_c a}$

- $a^{\log_a(x)} = x$

- $\ln(x) = \log_e(x)$

- $\log(x) = \log_{10}(x)$

- $\log_a(1) = 0$

- $\log_a(a) = 1$

In what way should we be careful when we use logarithmic property? Compare the following expressions to conclude that negative inputs are not allowed when we pull out exponents outside the logarithm.

$$\log_{2^2}\left((-4)^2\right) = \log_4(16) = 2$$

versus

$$\log_{2^2}\left((-4)^2\right) = \frac{2}{2}\log_2(-4) = \text{undefined}$$

6. Find the sum of solutions to $\log_9(x+3)^2 + \log_9(x-2)^2 = \log_{27}(2(x+3))^3$.

7. What is the largest integer less than $\log_2\frac{2}{1} + \log_2\frac{3}{2} + \cdots + \log_2\frac{2019}{2018} + \log_2\frac{2020}{2019}$?

John Napier, who is known to have invented logarithm, was a proponent of simplifying huge arithmetic down to something that we use with ease. His idea of $\log(a \cdot b) = \log(a) + \log(b)$ shows that multiplication turns into addition by using logarithm.

8. Given that $\log_2 \left(\dfrac{64^x}{16^y} \right)$ can be written as $ax + by$, find the value of a and of b.

Another application of logarithm, especially to the base 10, is computing the number of digits of a number. The idea begins with $\log(10) = 1.0$ and $\log(100) = 2$. Since the function is continuous, any real number betwen 10 and 100 has the integer digits of 1. Apply similar approach to the following question.

9. Given that $\log_{10} 2 \approx 0.3010$, how many digits are in 2^{2019}?

Notice that the range of $y = \log_a(x)$ is the set of all real numbers can be utilized in AM-GM inequality. Since $\log_a(b)$ and $\log_b(a)$ are real numbers, we can apply the famous inequality.

10. If $a \geq b > 1$, find the maximum value of $2 - (\log_a b + \log_b a)$.

Since $\log_a(b)$ is likely to be any real number, we use any letter to substitute this expression so that we may use some polynomial form to solve a given equation.

11. If $x > y > 0$, $\log_y x + \log_x y = \dfrac{17}{4}$, and $xy = 32$, then find the value of x.

Last application of logarithm may be found when we have variables in both base and exponent. For instance, if x^x is given, then we use logarithm to pull the exponent down as a constant multiple, i.e., $\log(x^x) = x \log(x)$.

12. Find the product of the positive roots of $\sqrt{10} \cdot x^{\log_{10} x} = x\sqrt{x}$.

13. If $\dfrac{\log_b a}{\log_c a} = \dfrac{20}{19}$, then $bc = c^{\frac{p}{q}}$ where p and q are relatively prime. Compute $p + q$.

Core Skill Practice 8

01.

If $\log(6) = a$ and $\log(5) = b$, express $\log(12)$ in terms of a and b.

02.

In how many points do the graphs of $y = \log_{10}(10x)$ and $y = 2\log_{10}(x)$ intersect?

03.

Find all the solutions of $x^{\log_2 x} = \dfrac{x^4}{8}$.

04.

If $x > 1, y > 1$, and $z = \dfrac{\log_y (\log_y x)}{\log_y x}$, then rewrite x^z in logarithmic form.

05.

Recall that

$$10^3 < 2^{10} < 2^{11} < 2^{12} < 2^{13} < 10^4$$

Find the largest rational a such that $a < \log_{10} 2$.

06.

Find the product of the positive roots of $x^{\log_{2020} x} = \dfrac{x^2}{\sqrt{2020}}$.

07.

Given that $\log_{10} 2 \approx 0.3010$, how many digits are in 5^{100}?

08.

Given that $\log_{4n}(8\sqrt{3}) = \log_{3n}(9) = k$, find the value of k.

09.

Suppose that p and q are positive real numbers for which

$$\log_4(p) = \log_{12}(q) = \log_{36}(4p + 3q) = K$$

What is the value of K?

10.

Suppose a and b are positive numbers for which

$$\log_4 a = \log_3 b = \log_5(a + b)$$

What is the value of $a \cdot b$?

Solution to Core Skill Practice 8

01.
Since $\log(2) + \log(3) = a$ and $\log(5) = 1 - \log(2) = b$, we conclude that
$\log(12) = 2\log(2) + \log(3) = a + 1 - b$.

02.
First, $y = 2\log_{10}(x) = \log_{10}(x^2)$. In order to find the intersection points, let's equalize the two functions, so $\log_{10}(10x) = \log_{10}(x^2)$. Since $y = \log_{10}(x)$ is a monotonically increasing function, it is $1 - 1$ function. Hence, $10x = x^2$. There are two values of x, i.e., $x = 0$ and $x = 10$. If $x = 0$, then $y = \log_{10}(x)$ is undefined. Therefore, $x = 10$ is the only solution. Therefore, there is one point of intersection.

03.

$$\log_2(x^{\log_2(x)}) = 4\log_2(x) - 3$$
$$(\log_2(x))^2 - 4\log_2(x) + 3 = 0$$
$$(\log_2(x) - 3)(\log_2(x) - 1) = 0$$
$$\log_2(x) = 3, 1$$
$$x = 2^3, 2^1$$
$$x = 8, 2$$

04.

$$z\log_y(x) = \log_y(\log_y(x))$$
$$\log_y(x^z) = \log_y(\log_y(x))$$
$$x^z = \log_y(z)$$

05. $\dfrac{3}{10} < \log_{10}(2)$.

06. The product of the positive roots are 2020^2.

07. There are 70 digits used for 5^{100}.

08. Since $(4n)^k = 8\sqrt{3}$ and $(3n)^k = 9$, we get $(4/3)^k = \dfrac{8}{3\sqrt{3}}$, Hence, $k = \dfrac{3}{2}$.

09. $K = \log_3(4)$.

10. $a \cdot b = 144$.

Can you compare 999^{1000} and 1000^{999}?

SOLUTION

#1. Notice that two numbers are both positive.

#2. Put "ln" infront of both expressions.

#3. $\ln(999^{1000}) = 1000 \ln(999)$ and $\ln(1000^{999}) = 999 \ln(1000)$.

#4. Look at $\dfrac{999}{1000}$ and $\dfrac{\ln(999)}{\ln(1000)}$.

#5. $\ln(999)$ and $\ln(1000)$ are closer to each other than 999 and 1000 are.

#6. Hence, $\dfrac{\ln(999)}{\ln(1000)}$ is closer to 1 than $\dfrac{999}{1000}$ is.

#7. Thus, we conclude that

$$\frac{999}{1000} < \frac{\ln(999)}{\ln(1000)}$$
$$999 \ln(1000) < 1000 \ln(999)$$
$$\ln(1000^{999}) < \ln(999^{1000})$$
$$1000^{999} < 999^{1000}$$

TOPIC
4

Trigonometric Ratio

Trigonometry is a combination of triangle and measurements. It all began in Greece and thrived in the 18th century. You may wonder how we got the words *sine* or *cosine* that appear mostly in Trigonometry. In medieval times, the study of trigonometry was stifled in Europe, but it witnessed its advancement in India. The study of trigonometry started with 'half-chord' and 'chord,' where half-chord is *jiva* in Sanskrit. When translated into Arabic, *jiva* is written as *jiba*, which simply turned into *jb*. As many people read it as *jayb*, which means 'inlet', Fibonacci translated Arabic works into Latin in the 13th century, naming *jayb* as *sinus*.

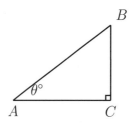

All the following functions, with different names, originated in Islam before the 13th century. Given a right triangle ABC and an angle θ, there are six trigonometric ratios.

(1) Sine : $\sin \theta = \dfrac{a}{c}$ (2) Cosine : $\cos \theta = \dfrac{b}{c}$ (3) Tangent : $\tan \theta = \dfrac{a}{b}$

(4) Cosecant : $\csc \theta = \dfrac{1}{\sin \theta}$ (5) Secant : $\sec \theta = \dfrac{1}{\cos \theta}$ (6) Cotangent : $\cot \theta = \dfrac{1}{\tan \theta}$

It is easy to check that Pythagorean theorem implies that $\sin^2(\theta) + \cos^2(\theta) = 1$. Also, there are special right triangles that will appear in Trigonometry. Two special triangles, $30° - 60° - 90°$ and $45° - 45° - 90°$, result in trigonometric ratios. Other than tangent values, sine value and cosine value are either $1/2$, $\sqrt{2}/2$, or $\sqrt{3}/2$. Notice that $\cos(45°)$ or $\sin(45°)$ are $1/\sqrt{2}$. If the denominator is rationalized, then the ratio turns into $\sqrt{2}/2$.

1. Given $\theta = 30°, 45°, 60°$, evaluate the following trigonometric ratio: $\sin(\theta)$, $\cos(\theta)$, and $\tan(\theta)$.

2.

(a) If $\tan(\theta) = 1$ for acute angle θ, find $\cos(\theta)$ and $\sin(\theta)$.

(b) If $\tan(\phi) = 3$, find $\cos(\phi)$ and $\sin(\phi)$ for an acute angle ϕ.

(c) If $\cos(\theta) = 1/2$ for acute angle θ, find $\sin(\theta)$ and $\tan(\theta)$.

(d) If $\sin(\theta) = x$ for acute angle θ, find $\cos(\theta)$ and $\tan(\theta)$.

Now, we will extend trigonometric ratio with triangles with more than acute angles. We need two things to extend on how we think about the ratio. First is the base angle (or reference angle), and the second is C.A.S.T sign rule.[1]

Reference angle is an *acute angle* formed between the x-axis and the rotated ray, also known as terminal arm. For instance, were $\sin(120°)$ to be computed, we use the fact that $60°$ is the reference angle. What we know from reference angle is that $|\sin(120°)| = |\sin(60°)|$. All we should know is the sign of $\sin(120°)$, determined by C.A.S.T rule.

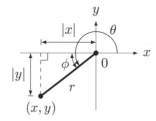

Try to compute $\cos(120°)$ using both reference angle and C.A.S.T rule. Its absolute value must be equal to $\cos(60°)$, while its sign is negative.

3. If $0° < \theta < 360°$, and θ has the reference angle (or base angle) of $45°$, find θ if

(a) it is in the 1st quadrant.   (b) it is in the 2nd quadrant.

(c) it is in the 3rd quadrant.   (d) it is in the 4th quadrant.

---

[1]In the first quadrant, **A**ll primary trigonometric ratios are positive. In the second quadrant, **S**ine ratios are positive, while the other primary trigonometric ratios are negative. In the third quadrant, **T**angent ratios are positive, while the other primary trigonometric ratios are negative. In the fourth quadrant, **C**osine ratios are positive, while other primary trigonometric ratios are negative.

By inspection, notice that

- $\cos(\bigcirc) = \cos(360° - \bigcirc)$ $\qquad$ Example) $\cos(120°) = \cos(240°)$.

- $\sin(\bigcirc) = \sin(180° - \bigcirc)$ $\qquad$ Example) $\sin(120°) = \sin(60°)$.

- $\tan(\bigcirc) = \tan(180 + \bigcirc)$ $\qquad$ Example) $\tan(210°) = \tan(30°)$.

- $\cos(\bigcirc) = \sin(90° - \bigcirc)$ $\qquad$ Example) $\cos(30°) = \sin(60°)$.

4. Without using a calculator, evaluate the following expressions. If the value is undefined, write *undefined*.

(a) $\dfrac{\cos(135°)}{\sin(120°) + \cos(150°)}$

(b) $\tan(135°) - \tan(225°)\tan(-45°)$

(c) $\dfrac{\sin(30°)}{\cos(60°)}$

(d) $\sec(135°)$

(e) $\tan(210°)$

(f) $\cos(330°) + \sin(300°)$

5. If $\sin(\theta) = \dfrac{1}{2}$ for acute angle θ, find the value of the following expressions.

(a) $\cos(90° - \theta)\sin(\theta)$

(b) $\cot(\theta)\tan(90° - \theta)$

(c) $\cos^2(\theta) + \sin^2(\theta)$

(d) $1 + \tan^2(\theta)$

(e) $1 + \cot^2(\theta)$

(f) $\sec^2(90° - \theta) - \tan^2(90° - \theta)$

6.

(a) If $\cos(\theta) = \dfrac{1}{2}$ and $270° < \theta < 360°$, find $\sin(\theta) + \tan(\theta)$.

(b) If $\cos(\theta) = -\dfrac{2}{3}$ and $180° < \theta < 270°$, find $\tan(\theta)$ and $\sin(\theta)$.

(c) If $\sin(\theta) = -\dfrac{4}{5}$ and $180° < \theta < 270°$, find $\cot(\theta)$ and $\sec(\theta)$.

(d) If $\tan(\theta) = \dfrac{1}{3}$ and $180° < \theta < 270°$, find $\sec(\theta)$ and $\csc(\theta)$.

7. In right triangle ABC with $\angle B = 90°$, we have

$$4\sin A = 7\cos A$$

What is $\sin A$?

8. Find the integer value of n, for $0 \le n \le 180$, such that $\cos(n°) = \cos(789°)$.

9. Find the integer value of n, for $-90 \le n \le 90$, such that $\sin(n°) = \sin(543°)$.

Core Skill Practice 9

01.
Evaluate each of the following expressions.

(a) $\sin(150°)$

(b) $\cos(300°)$

(c) $\tan(225°)$

(d) $\cot(135°)$

(e) $\sec(-45°)$

(f) $\csc(-30°)$

02.
For how many values of θ such that $0° \leq \theta \leq 720°$ is $\cos(\theta) = \dfrac{\sqrt{3}}{2}$?

03.
Find the principal angle for the given reference angle and quadrant.

(a) 30°, Quadrant II

(b) 60°, Quadrant IV

(c) 25°, Quadrant III

04.
Find the quadrant in which the terminal arm of θ lies, from the given information.

(a) $\sin(\theta) > 0, \cos(\theta) < 0$

(b) $\tan(\theta) < 0, \cos(\theta) > 0$

(c) $\sec(\theta) < 0, \csc(\theta) > 0$

05.
Find the other two primary trigonometric ratios with the given ratio and quadrant, denoted by Q1, Q2, Q3, and Q4, in the following parts.

(a) $\sin(\theta) = \dfrac{1}{a}$, Q2

(b) $\cos(\theta) = \dfrac{n}{m}$, Q4

(c) $\tan(\theta) = \dfrac{p}{q}$, Q3

06.
Find the exact value of each ratio. Do not use a calculator.

(a) $\csc(150°)$

(b) $\sec(210°)$

(c) $\cot(135°)$

07.

Solve for θ where $0° \leq \theta < 360°$. Show your work.

(a) $\sin(\theta) = \dfrac{1}{2}$

(b) $\cos(\theta) = -\dfrac{1}{\sqrt{2}}$

(c) $\tan(\theta) = -\dfrac{\sqrt{3}}{3}$

(d) $\cos(\theta) = -\dfrac{1}{2}$

(e) $\cos(\theta) = \dfrac{\sqrt{3}}{2}$

(f) $\sin(\theta) = -1$

01.

(a) $\sin(150°) = \dfrac{1}{2}$.

(b) $\cos(300°) = \dfrac{1}{2}$.

(c) $\tan(225°) = 1$.

(d) $\cot(135°) = -1$.

(e) $\sec(-45°) = \sqrt{2}$.

02. There are four values of θ.

03.

(a) $150°$

(b) $300°$

(c) $205°$

04.

(a) 2nd quadrant.

(b) 4th quadrant.

(c) 2nd quadrant.

05.

(a) $\cos(\theta) = -\dfrac{\sqrt{a^2 - 1}}{a}$, $\tan(\theta) = -\dfrac{1}{\sqrt{a^2 - 1}}$.

(b) $\sin(\theta) = -\dfrac{\sqrt{m^2 - n^2}}{|m|}$, $\cos(\theta) = -\dfrac{\sqrt{m^2 - n^2}}{|n|}$.

(c) $\cos(\theta) = -\dfrac{|q|}{\sqrt{p^2 + q^2}}$, $\sin(\theta) = -\dfrac{|p|}{\sqrt{p^2 + q^2}}$.

06.

(a) $\csc(150°) = 2$.

(b) $\sec(210°) = -\dfrac{2\sqrt{3}}{3}$.

(c) $\cot(135°) = -1$.

07.

(a) $\theta = 30°, 150°$.

(b) $\theta = 135°, 225°$.

(c) $\theta = 150°, 330°$.

(d) $\theta = 120°, 240°$.

(e) $\theta = 30°, 330°$.

(f) $\theta = 270°$.

TOPIC
5

Trigonometric Functions

In this section, we mix-use π as $180°$, converting degree units to radians and vice versa. First, we need a unit circle to begin with. Understanding the unit circle is most important in graphing trigonometric functions.

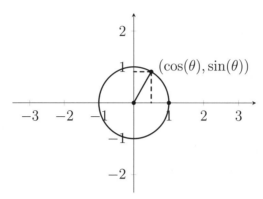

The figure above is for $r = 1$, Otherwise, $x = r\cos(\theta)$ and $y = r\sin(\theta)$. Here, assuming $r = 1$, the point on the unit circle has the coordinates corresponding to $\cos(\theta)$ and $\sin(\theta)$ where θ is the amount of counterclockwise rotation from the initial point $(1, 0)$ on the x-axis. We consider θ positive if the angle rotated is counterclockwise. Otherwise, θ is negative.

- The x-coordinate of the endpoint of the terminal arm : $\cos(\theta)$.

- The y-coordinate of the endpoint of the terminal arm : $\sin(\theta)$.

- The slope of the terminal arm : $\tan(\theta)$.

Hence, we get the following analyses directly from the above unit circle.

- $-1 \le \cos(\theta) \le 1$ and $-1 \le \sin(\theta) \le 1$.

- $\tan(\theta)$ is undefined at $\theta = \cdots, \dfrac{-\pi}{2}, \dfrac{\pi}{2}, \dfrac{3\pi}{2}, \dfrac{5\pi}{2}, \cdots$.

- $\cos(\theta)$ and $\sin(\theta)$ show repetitive values since the terminal arm returns to the starting point $(1, 0)$.

- $\tan(\theta)$ results in repetitive values every halfcircle. Think about why it happens, and connect with the slope of a line containing the terminal arm.

Instead of using the angle in degree measure, we may use *radian* measure because radian measures length. In Geometry, we learned how to compute the arc length of a given circle. Now, we will use an associated arc length, instead of degree measure. For instance, $90°$ corresponds to $\dfrac{\pi}{2}$. Likewise, $180°$ corresponds to π, coming from the unit circle.

Example

Convert $150°$ into radian.

Solution

$150° \times \dfrac{\pi \text{ radian}}{180°} = \dfrac{5\pi}{6}$.

Let's first define two functions that explain about sine and cosine functions, both of which are knwon as sinusoidal functions.

Periodic Function

Its graph portions are continuously copy-and-pasted.

$$f(x + k) = f(x)$$

If $\cdots = f(3) = f(5) = f(7) = \cdots$, then we say $f(x)$ is periodic with the period of 2. Given the graphs of trigonometric functions, we can find the period by looking at the horizontal distance between two adjacent maxima(or minima).

Bounded Function

Its graph has both maximum and minimum. In this case, $f(x) = \cos(x)$ and $f(x) = \sin(x)$ are bounded function such that
$$-1 \leq \cos(x)(\text{or } \sin(x)) \leq 1$$

Example

Given $y = f(x) = 2 + 3\sin(x)$, compute its range.

Solution
The range of the function is given by $[-1, 5]$ since

$$-1 \leq \sin(x) \leq 1$$
$$-3 \leq 3\sin(x) \leq 3$$
$$-3 + 2 \leq 2 + 3\sin(x) \leq 3 + 2$$

1. If $f(x + 2) = f(x)$ and $f(1) = \pi$, find

$$\frac{f(2021)f(2019)f(2017)\cdots f(1011)}{f(1009)f(1007)\cdots f(1)}$$

2. Which of the following is the closest integer to $\dfrac{100}{\pi}$?

(A) 31
(B) 32
(C) 33
(D) 34
(E) 35

Sine function

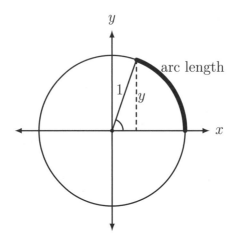

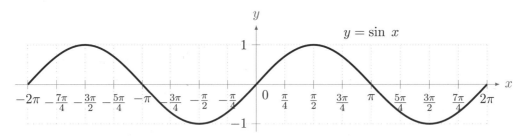

Given $y = \sin(x)$, the input is the arc length on the unit circle, since $l = r\theta = \theta$ for $r = 1$, whereas the output is the y-coordinate of the endpoint of the terminal arm on the unit circle. Look at the graph, which is bounded and periodic.

- Domain(the set of x-values) is $\mathbb{R}$.

- Range(the set of y-values) is $[-1, 1]$.

- Period is 2π.

- This is an odd function, satisfying $\sin(-x) = -\sin(x)$.

Example

Draw a vertical line at $x = \dfrac{\pi}{2}$ and check that $y = \sin(x)$ is symmetric with respect to the line.

Solution
Since $\sin(x) = \sin(\pi - x)$, the graph must be symmetric with respect to the vertical line at the average of x and $\pi - x$. Hence, $x = \dfrac{\pi}{2}$ must be a line of symmetry.

3. Find the number of real solutions to $f(g(x)) = 0$ where $f(x) = \sin(x)$ and $g(x) = \dfrac{1}{x}$, where $x \in (0.001, 0.01)$.

4. For $f(x) = \sin(2\pi x)$ defined on $x \in [-2, 2]$, find the number of x-intercepts of $f(f(x))$.

Cosine Function

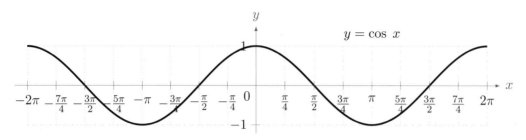

Inputs are the arc lengths traced by moving the arms around the unit circle, usually in counter-clockwise direction, and outputs are the x-coordinates of the endpoint of the terminal arm on the circle. Like sine graph, the graph is bounded and periodic.

- Domain(the set of x-values) is $\mathbb{R}$.

- Range(the set of y-values) is $[-1, 1]$.

- Period is 2π.

- This is an even function, satisfying $\cos(-x) = \cos(x)$.

Given $f(x) = \cos(2\pi - x)$, the function is periodic function with the period of 2π. Therefore, $f(x) = f(x + 2\pi)$. Hence, $f(x + 2\pi) = \cos(2\pi - (x + 2\pi)) = \cos(-x) = \cos(x)$. We can simplify expressions using periodicity and odd/evenness of the function. The following diagram is the graph of $f(x) = \cos(2\pi - x)$, which is not different from that of $y = \cos(x)$.

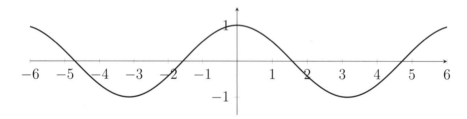

Example

Compute the range of $y = 2 + 3|\cos(x + \pi)|$.

Solution
Since $-1 \le \cos(x + \pi) \le 1$, $0 \le |\cos(x + \pi)| \le 1$. Hence,

$$0 \le 3|\cos(x + \pi)| \le 3$$
$$2 \le 2 + 3|\cos(x + \pi)| \le 5$$

The range of the function is $[2, 5]$.

5. Find the number of real solutions to $f(g(x)) = 1$ where $f(x) = \cos(x)$ and $g(x) = \dfrac{1}{x}$, where $x \in (0.001, 0.01)$.

6. For $f(x) = |\cos(\pi x)|$, find the number of real solutions to $f(f(x)) = \dfrac{1}{2}$ where $0 \leq x \leq 2$.

Tangent Function

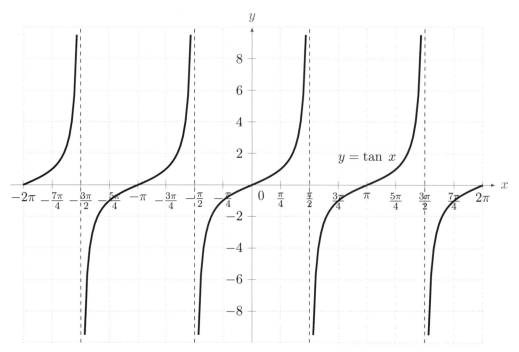

$$y = \tan x$$

Inputs are the arc lengths traced by moving the arms around the unit circle, usually in counter-clockwise direction, and outputs are the slopes of the end-tip of rays on a unit circle. This function is not bounded, but periodic.

- Domain is $\mathbb{R} \setminus \{n\pi + \frac{\pi}{2}\}$ or $\{x \,|\, x \in \mathbb{R}, x \neq \frac{2n+1}{2}\pi \,(n \in \mathbb{Z})\}$.

- Range is $\mathbb{R}$.

- Period is π.

- It is an odd function, satisfying $\tan(-x) = -\tan(x)$.

The graph of $y = \tan(x)$ has vertical asymptotes because of $\tan(x) = \dfrac{\sin(x)}{\cos(x)}$. Since denominator must be nonzero, we get $\cos(x) \neq 0$.

Example

Find the vertical asymptotes of $y = \tan(-x)$.

Solution

The graph of $y = \tan(-x) = -\tan(x)$ is simply the reflection of $y = \tan(x)$ about the x-axis. Hence, the vertical asymptotes don't change. They are
$$\frac{\pm\pi}{2}, \frac{\pm 3\pi}{2}, \frac{\pm 5\pi}{2}, \ldots.$$

7. If $\sin(\cos(\pi x)) = 1/2$ for $x \in [0, 1]$, find the number of real x's.

8. Given $0 \leq x \leq \dfrac{\pi}{4}$, find the range of values of $2 + 3\tan(x)$.

9. Sketch the following graphs on the following xy-plane with proper labels.

(a) $y = \sin |x|$

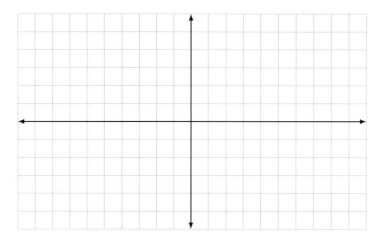

(b) $y = |\tan(x)|$

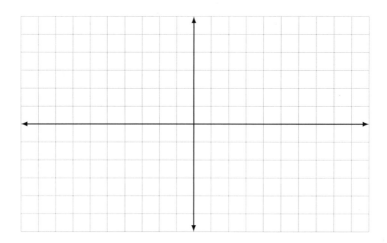

10. Sketch the graph of $y = \lfloor \cos(\pi x) \rfloor$ for $0 \le x \le 4$.

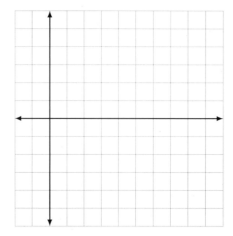

11. Which of the following is the period of the function $\sin(\frac{\pi}{2}x) + \cos(\frac{\pi}{3}x) + \tan(\pi x)$?

(A) 2
(B) 4
(C) 6
(D) 10
(E) 12

12. If $f(x) = \sqrt{1 + \sin(x)} + \sqrt{1 - \sin(x)}$ has the period of p, which of the following is equal to the value of $f(p)$?

(A) π
(B) 2π
(C) $\sqrt{2}$
(D) 2
(E) 1

Sinusoidal functions refer to sine functions or cosine functions, which typically can be written as

$$y = A\sin(B(x - C)) + D \text{ or } y = A\cos(B(x - C)) + D$$

- Amplitude $= |A|$, defined as half the difference between the maximum and minimum value.

- Period $= \dfrac{2\pi}{|B|}$, which is the amount of curve that can trace out the whole graphs by copy-and pasting.

- Phase Shift $= C$ (units to the right), which is equal to horizontal translation.

- Vertical Shift $= D$ (units up), which is equal to vertical translation.

Look at the graph of $y = \sin(2x)$ in the figure below and see how it is horizontally shrunk by a factor of 2 compared to that of $y = \sin(x)$.

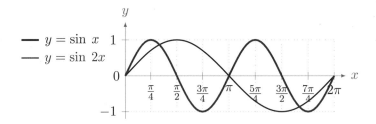

Example

If amplitude is 2, period is $\dfrac{\pi}{3}$ and the phase shift is 2, find the sine function that satisfies this property.

> **Solution**
> $y = f(x) = 2\sin(6(x - 2))$.

Also, look at the graph of $y = 3\cos(2x)$ below, which changes the amplitude and period at the same time.

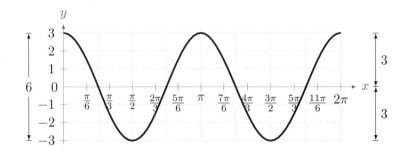

13. If the positive x-coordinates of the first three points of intersection between the graphs of $f(x) = 4\sin(x)$ and $y = 3$ are a, b, and c, which of the following is the value of $f(a+b+c)$?

(A) -4
(B) 4
(C) 3
(D) -3
(E) 0

14. Let $f(x) = \begin{cases} 1 & \text{if } 0 \le x \\ 0 & \text{if } x < 0 \end{cases}$, then find the number of intersections between

$g(x) = f(x)f(\pi - x)\sin(2x)$ and $h(x) = \dfrac{1}{3}$.

There are three reciprocal trigonometric functions known as cosecant function, secant function, and cotangent function.

$$\csc(x) = \frac{1}{\sin(x)} \qquad\qquad \sec(x) = \frac{1}{\cos(x)} \qquad\qquad \cot(x) = \frac{1}{\tan(x)} = \frac{\cos(x)}{\sin(x)}$$

Secant Function

Secant function is a reciprocal of cosine function. Mathematical expression for the secant function is

$$f(x) = \sec(x) = \frac{1}{\cos(x)}$$

In this case, the domain must be carefully considered. In other words, the values of x such that $\cos(x) = 0$ should be eliminated from our domain because those values make the function undefined.

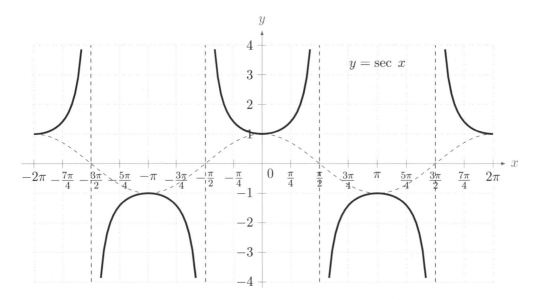

The following properties are satisfied.

1. Domain = the set of all values of x except $x = \pm\dfrac{\pi}{2}, \pm\dfrac{3\pi}{2}, \cdots$.

2. Range = the set of all values of y such that $y \le -1$ or $y \ge 1$.

3. There are infinitely many vertical asymptotes $x = \pm\dfrac{\pi}{2}, \pm\dfrac{3\pi}{2}, \cdots$.

4. Period = same as that of cosine.

5. Discontinuous function, which means the graph is disconnected.

6. This is even function, whose graph is symmetric with respect to the y-axis.

Cosecant Function

Cosecant function is a reciprocal of sine function, i.e.,

$$f(x) = \csc(x) = \frac{1}{\sin(x)}$$

The following properties are satisfied.

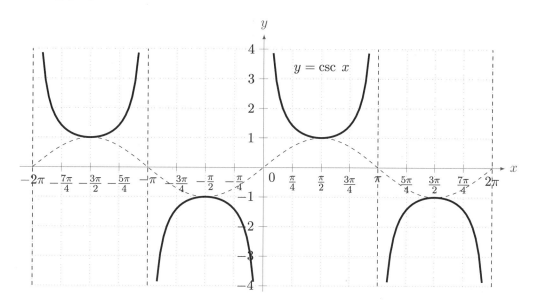

1. Domain = the set of all values of x except $x = \pm\pi, \pm2\pi, \pm3\pi \cdots$.

2. Range = the set of all values of y such that $y \leq -1$ or $y \geq 1$.

3. There are infinitely many vertical asymptotes $x = 0, \pm\pi, \pm2\pi, \pm3\pi, \cdots$.

4. Period = same as that of sine.

5. Discontinuous function, which means its graph is disconnected, due to undefined x-inputs.

6. This is odd function, whose graph is symmetric with respect to the origin.

There are two rules to satisfy when we draw the reciprocal function of $f(x)$.

- The sign of y-value does not change. If $f(x)$ is positive, then $\dfrac{1}{f(x)}$ is also positive.

- x-intercepts become vertical asymptotes, and vice versa.

- End-behavior is also switched from $\pm\infty$ to horizontal asymptote $y = 0$, and vice versa.

It is easy to check that the graph of cosecant function follows the rules stated above when we simply use the reciprocal of cosine function. It is easy to remember how we derive the graph of cosecant function by *flipping* each mound-like portion of cosine function to the opposite direction.

Cotangent Function

Cotangent function is a reciprocal of tangent function. Mathematical expression for the cotangent function is

$$f(x) = \cot(x) = \frac{1}{\tan(x)}$$

Similar to $\csc(x)$, we do not allow $\sin(x) = 0$ for x values. Hence, the domain for cotangent function is equivalent to that of cosecant function. The following lists are the properties of cotangent function.

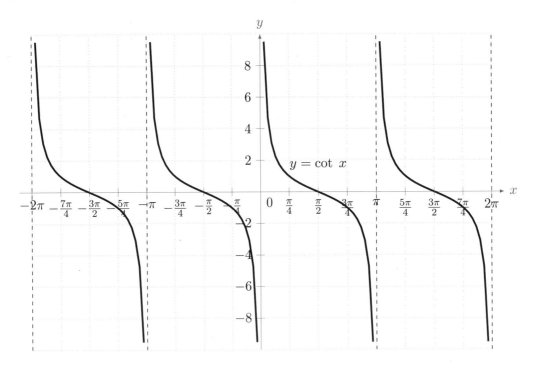

1. Domain = the set of all values of x except $x = 0, \pm\pi, \pm2\pi, \pm3\pi \cdots$.

2. Range = $\mathbb{R}$.

3. There are infinitely many vertical asymptotes $x = 0, \pm\pi, \pm2\pi, \pm3\pi, \cdots$.

4. Period = same as that of tangent.

5. Discontinuous function, which means the graph is disconnected.

6. This is odd function, whose graph is symmetric with respect to the origin.

Unlike the previous two graphs, the graph of cotangent function is not that easy to remember in terms of tangent function. The quickest way to remember the graph is that tangent function goes up in one period, whereas cotangent function goes down in the corresponding one period.

15. The following is the graph of $y = A\csc(Bx)$ for some positive real A and B. Find $A \cdot B$.

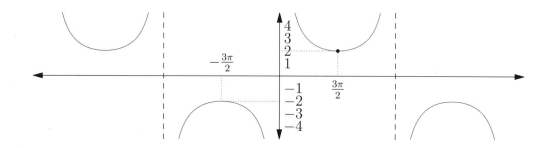

Now, it is about time for us to see the graph behavior of $\sin(x)$, $\tan(x)$ and x. Though we will not prove it because it requires some Calculus-level knowledge, we would like to see how the graphs of $\sin(x)$, $\tan(x)$ and x resemble one another near $x = 0$.

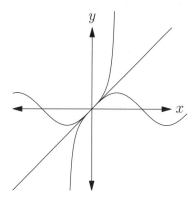

Around $x \approx 0$, the graphs of $y = \sin(x)$, $y = \tan(x)$ and $y = x$ are almost indistinguishable, which can be non-mathematically written as

$$\sin(x) \approx \tan(x) \approx x \text{ if } x \text{ is small}$$

This requires the concept of limit, which can be mathematically written as

$$\lim_{x \to 0} \frac{\sin(x)}{x} = \lim_{x \to 0} \frac{\tan(x)}{x} = \lim_{x \to 0} \frac{\sin(x)}{\tan(x)} = 1$$

16. The graph of $y = |\cot(2x)|$ defined on x in $(0, \frac{\pi}{2})$ is above the graph of $y = k|x - \frac{\pi}{4}|$, except at $x = \frac{\pi}{4}$. Find the maximum value of k.

We will have a look at maximum and minimum of trigonometric functions, which involves

- t-substitution for algebraic expression where $\sin(x) = t$ or $\cos(x) = t$.

- vertex form of a quadratic function $y = a(x - h)^2 + t$.

In fact, we need to identify the range and substitute a given trigonometric function to do algebra.

17. Find the maximum and minimum of the following trigonometric functions.

(a) $y = |\cos(x) - 2| + 2$

(b) $y = 3 - |2 - \sin(x)|$

18. Find the maximum and minimum of the following trigonometric functions.

(a) $y = \cos^2(x) - 5\cos(x) + 4$ where $0 \leq x \leq \pi$.

(b) $y = \tan^2(x) + \tan(x)$ where $-\dfrac{\pi}{4} \leq x \leq \dfrac{\pi}{4}$.

19. For $0 \le x \le 2\pi$, if the maximum value of $y = -\cos^2(x) + 2a\cos(x)$ is 3, then find the positive value of a.

20. Find the difference between maximum and minimum of $y = \dfrac{2\sin(x)}{\sin(x) + 4}$.

Solving Trigonometric Equations

Solving trigonometric equations usually requires the following steps, using some knowledge about sine, cosine, and tangent properties. Here is the general rule for solving trigonometric equations.

(1) Find the reference angle.

(2) Remember that $\sin(\theta)$ is the y-coordinate, $\cos(\theta)$ is the x-coordinate and $\tan(\theta)$ is the gradient of the terminal arm. Draw the terminal arms in the coordinate plane with C.A.S.T.

(3) Find the correct values that fit in the range of x-values.

21. Solve the following trigonometric equations.

(a) $\cos(2x) = \sin(2x)$ for $0° \leq x \leq 360°$.

(b) $\cot^2(x) + 3\csc(x) = 3$ for $0° \leq x \leq 180°$.

22. Solve $\sin(x + \frac{\pi}{4}) = -\frac{1}{2}$ for $0 \le x \le 2\pi$ radians, giving each answer as a multiple of π.

23. Solve $2\cos(x)\cot(x) + 1 = \cot(x) + 2\cos(x)$ for $0 < x < \pi$.

Core Skill Practice 10

01.

Find the period, amplitude, and phase shift of $f(x) = -3\cos(2x + \pi) + 4$.

02.

Find the period of $y = 2\sin(x) - \tan(x)$.

03.

What is the period of $g(x) = -2\tan(3x - \pi)$?

04.

Find the asymptotes of the graph of $y = \cot 2x$.

05.

Find all angles between 0 and 2π, which satisfy the following equations.

(a) $\cos(2x) = \dfrac{1}{2}$

(b) $\tan(x - \dfrac{\pi}{3}) = \dfrac{\sqrt{3}}{3}$

(c) $\sin(\dfrac{x}{2}) = \dfrac{1}{2}$

06.

If θ is an obtuse angle and $2\tan^2(\theta) - 5\sec(\theta) = 10$, find the value of $\tan(\theta)$ without using a calculator.

07.

If $\sin(x) - \cos(x) = \dfrac{1}{4}$, evaluate

(a) $\sin(x)\cos(x)$

(b) $\sec(x) - \csc(x)$

08.

If $\sin(x) + \sin(y) = a$ and $\cos(x) + \cos(y) = a$, where $a \neq 0$, find $\sin(x) + \cos(x)$ in terms of a.

09.

If $x = \sin(\theta) - 2\cos(\theta)$ and $y = 2\sin(\theta) + \cos(\theta)$, find a relation between x and y independent of θ.

10.

Solve $(\sin(x))^{\sin(2x)-\cos(2x)} = 1$ for $0 \leq x \leq 2\pi$.

💡 Solution to Core Skill Practice 10

01. Period is π, amplitude is 3, phase shift is $\dfrac{\pi}{2}$(left), and vertical shift is 4(up).

02. Period is 2π.

03. Period is $\dfrac{\pi}{3}$.

04. Vertical asymptotes are $x = 0, \pm\dfrac{\pi}{2}, \pm\pi, \pm\dfrac{3\pi}{2}, \cdots$.

05.
(a) $x = \dfrac{\pi}{6}, \dfrac{5\pi}{6}, \dfrac{7\pi}{6}, \dfrac{11\pi}{6}$.

(b) $x = \dfrac{\pi}{2}, \dfrac{3\pi}{2}$.

(c) $x = \dfrac{\pi}{3}, \dfrac{5\pi}{3}$.

06. $\tan(\theta) = -\dfrac{\sqrt{5}}{2}$.

07.
(a) $\sin(x)\cos(x) = \dfrac{15}{32}$.

(b) $\sec(x) - \csc(x) = \dfrac{8}{15}$.

08. $\sin(x) + \cos(x) = a$.

09. $x^2 + y^2 = 5$.

10. $x = \dfrac{\pi}{8}, \dfrac{\pi}{2}, \dfrac{5\pi}{8}, \dfrac{9\pi}{8}, \dfrac{13\pi}{8}$.

When we look for inverse function, the original function must be one-to-one function. Trigonometric functions are not one-to-one, so we must <u>restrict our domain</u> to come up with inverse functions.

$$y = \sin(x) : \left[-\frac{\pi}{2}, \frac{\pi}{2}\right] \qquad y = \cos(x) : [0, \pi] \qquad y = \tan(x) : \left(-\frac{\pi}{2}, \frac{\pi}{2}\right)$$

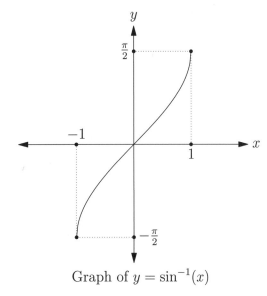

Graph of $y = \sin^{-1}(x)$

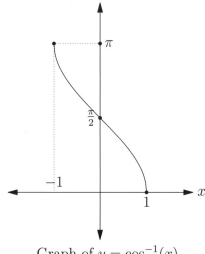

Graph of $y = \cos^{-1}(x)$

We should remember the properties of inverse functions : $f(f^{-1}(x)) = x$ and $f^{-1}(f(x)) = x$. Another important fact to remember is that the domain must be carefully identified.

Example

Find the exact value of $\sin^{-1}(\sin(\frac{\pi}{4}))$.

Solution

Since $\sin(\frac{\pi}{4}) = \frac{\sqrt{2}}{2}$, we get $\sin^{-1}(\frac{\sqrt{2}}{2}) = \frac{\pi}{4}$.

24. Find the exact value of $\sin^{-1}(\sin(\frac{2\pi}{3}))$.

On the other hand, the composition in different order results in different outcomes.

Compute $\sin(\sin^{-1}(0.3))$.

Solution
Since 0.3 is between -1 and 1, then $\sin(\sin^{-1}(0.3)) = 0.3$

25. Find the exact value of $\sin(\sin^{-1}(1.1))$.

26. Evaluate each of the following expressions.

(a) $\arcsin(0)$ 　　　　(b) $\cos^{-1}(\frac{\sqrt{3}}{2})$ 　　　　(c) $\arctan(1)$

27. Evaluate each of the following expressions without a calculator.

(a) $\arcsin(\sin(39\pi/5))$

(b) $\arcsin(\sin(37\pi/5))$

(c) $\sin(\arccos(0.6))$

(d) $\sin(\arccos(-0.6))$

The arc-secant, arc-cosecant, and arc-cotangent functions are defined as

- $y = \sec^{-1}(x)$: the range is $[0, \pi]$ where $y \neq \dfrac{\pi}{2}$.

- $y = \csc^{-1}(x)$: the range is $\left[-\dfrac{\pi}{2}, \dfrac{\pi}{2}\right]$ where $y \neq 0$.

- $y = \cot^{-1}(x)$: the range is $(0, \pi)$.

The reason why $f(x) = \sec^{-1}(x)$ and $f(x) = \csc^{-1}(x)$ has an undefined y-value is because it is a reciprocal function.

Example

Find the exact value of $\sec^{-1}(2)$.

> **Solution**
> Let $\theta = \sec^{-1}(2)$. Then, we are looking for the angle θ where $0 \leq \theta \leq \pi$ except $\theta = \dfrac{\pi}{2}$, whose secant value equals 2, which means $\cos(\theta) = \dfrac{1}{2}$. Therefore, $\theta = \dfrac{\pi}{3}$, which is the only angle that works in this interval. Therefore, $\sec^{-1}(2) = \dfrac{\pi}{3}$.

28. Find the exact values of the following inverse trigonometric expressions, without using a calculator.

(a) $\cot^{-1}(1)$ (b) $\cot^{-1}(-1)$

29.

(a) Find the value of $\arccos(x) + \arcsin(x)$ for well-defined values of x.

(b) Find the sum of $\arccos(\cos(7)) + \arcsin(\sin(5))$, in radians.

30. Find the area bounded by the graph of $y = \arccos(\sin x)$ and the x-axis on the interval $-\dfrac{3\pi}{2} \leq x \leq \dfrac{\pi}{2}$.

Core Skill Practice 11

01.

Find the exact value of $\sin\left(\tan^{-1}\left(\dfrac{1}{3}\right)\right)$.

02.

Find the exact value of $\cos\left(\arcsin\left(-\dfrac{1}{2}\right)\right)$.

03.

Find the exact value of $\tan\left(\arccos\left(-\dfrac{2}{3}\right)\right)$.

04.

Find the exact value of $\csc^{-1}(2\sqrt{3}/3)$.

05.

(a) Write $\cos(\tan^{-1}(x))$ as an algebraic expression in terms of x.

(b) Write $\tan(\cos^{-1}(x))$ as an algebraic expression in terms of x.

06.

Let $f(x) = \sin(x)$ for $\left[-\frac{\pi}{2}, \frac{\pi}{2}\right]$, $g(x) = \cos(x)$ for $[0, \pi]$. Find $g\left(f^{-1}\left(\frac{5}{13}\right)\right)$.

Solution to Core Skill Practice 11

01. $\dfrac{\sqrt{10}}{10}$

02. $\dfrac{\sqrt{3}}{2}$

03. $-\dfrac{\sqrt{5}}{2}$

04. $\dfrac{\pi}{3}$

05.

(a) $\sqrt{\dfrac{1}{x^2+1}}$

(b) $\dfrac{\sqrt{1-x^2}}{x}$

06. $\dfrac{12}{13}$

#1. Do you notice that $\sin(0.001)$ is extremely similar to 0.001?

#2. $\tan(0.001)$ is extremely similar to $\sin(0.001)$ as well.

#3. $e^{0.001} - 1$ is also extremely similar to 0.001, though slightly bigger.

#4. $\ln(1.001)$ is also close to 0.001, though slightly smaller.

#5. Later in calculus, we learn that $(\sin^{-1}(x))' = \dfrac{1}{\sqrt{1 - x^2}} > 0$, which implies that the graph of $y = \sin^{-1}(x)$ goes up.

#6. We will also learn that $(\cos^{-1}(x))' = \dfrac{-1}{\sqrt{1 - x^2}} < 0$, which implies that the graph of $y = \cos^{-1}(x)$ goes down.

TOPIC
6

Trigonometric Identities

This section is dedicated to the father of modern algebra and trigonometry, Francois Viete. Not only did he come up with Viete's formula, but he also came up with multiple-angle formula and trigonometric solution of a cubic equation.

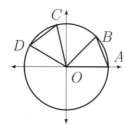

In the diagram, if $\angle AOD = \alpha$ and $\angle AOC = \beta$, let $\theta = \alpha - \beta$, which is $\angle COD$. Rotate $\overline{OA}$ counterclockwise by θ to get $\overline{OB}$. Then, we could write down the coordinate of the points on the unit circle as

$$A = (1, 0)$$
$$B = (\cos(\theta), \sin(\theta))$$
$$C = (\cos(\beta), \sin(\beta))$$
$$D = (\cos(\alpha), \sin(\alpha))$$

where $AB = CD$. Applying distance formula, we get

$$(1 - \cos(\theta))^2 + \sin^2(\theta) = (\cos(\alpha) - \cos(\beta))^2 + (\sin(\alpha) - \sin(\beta))^2$$
$$1 - 2\cos(\theta) + \cos^2(\theta) + \sin^2(\theta) = 2 - 2\cos(\alpha)\cos(\beta) - 2\sin(\alpha)\sin(\beta)$$
$$\cos(\theta) = \cos(\alpha)\cos(\beta) + \sin(\alpha)\sin(\beta)$$
$$\cos(\alpha - \beta) = \cos(\alpha)\cos(\beta) + \sin(\alpha)\sin(\beta)$$

Thus, we get the difference formula. Hence, if we substitute $-\beta$ instead of β, we get

$$\cos(\alpha - (-\beta)) = \cos(\alpha)\cos(\beta) + \sin(\alpha)\sin(-\beta)$$
$$\cos(\alpha + \beta) = \cos(\alpha)\cos(\beta) - \sin(\alpha)\sin(\beta)$$

If we use the complementary angle theorem, i.e., $\cos(\frac{\pi}{2} - (\alpha + \beta)) = \sin(\alpha + \beta)$, then we get

$$\sin(\alpha + \beta) = \cos(\frac{\pi}{2} - \alpha - \beta)$$
$$= \cos\left((\frac{\pi}{2} - \alpha) - \beta\right)$$
$$= \cos(\frac{\pi}{2} - \alpha)\cos(\beta) + \sin(\frac{\pi}{2} - \alpha)\sin(\beta)$$
$$= \sin(\alpha)\cos(\beta) + \cos(\alpha)\sin(\beta)$$

Hence, we also get the sum formula for sine as well.

$$\sin(\alpha - \beta) = \sin(\alpha)\cos(-\beta) + \cos(\alpha)\sin(-\beta)$$
$$= \sin(\alpha)\cos(\beta) - \cos(\alpha)\sin(\beta)$$

- $\cos(-x) = \cos(x)$, $\sin(-x) = -\sin(x)$, and $\tan(-x) = -\tan(x)$
  Both sine and tangent are odd functions, whereas cosine is even.

- $\cos^2(x) + \sin^2(x) = 1$, $\tan^2(x) + 1 = \sec^2(x)$, and $\cot^2(x) + 1 = \csc^2(x)$
  Apply Pythagorean theorem to a right triangle on the unit circle to get the first identity, and divide it by either the square of cosine or sine to derive the second and third identity.

- $\cos(x \pm y) = \cos(x)\cos(y) \mp \sin(x)\sin(y)$
  Abu al-Wafa, a Persian astronomer in the 10th century, gave another proof that showed resemblence of what we know today. It is marvelous to see that this idea was solidly used in 900's.

- $\sin(x \pm y) = \sin(x)\cos(y) \pm \cos(x)\sin(y)$
  As shown in the following figure, the proof uses the segment addition postulate.

- $\cos(\frac{\pi}{2} - x) = \sin(x)$ or vice versa
  A direct application of previous two bullet points.

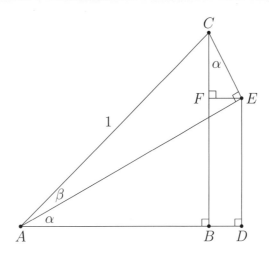

- $AE = \cos(\beta)$, $CE = \sin(\beta)$.

- $AB = \cos(\alpha + \beta)$, $CB = \sin(\alpha + \beta)$

- $DE = FB = \cos(\beta)\sin(\alpha)$

- $AD = \cos(\beta)\cos(\alpha)$

- $BD = \sin(\beta)\sin(\alpha)$, $CF = \sin(\beta)\cos(\alpha)$

Segment addition postulate implies that $AB + BD = AD$ and $CF + FB = CB$. Hence,

$$\cos(\alpha + \beta) = \cos(\alpha)\cos(\beta) - \sin(\alpha)\sin(\beta)$$
$$\sin(\alpha + \beta) = \sin(\alpha)\cos(\beta) + \cos(\alpha)\sin(\beta)$$

1. Prove the following identities.

(a)
$$\frac{1}{1-\cos\theta} + \frac{1}{1+\cos\theta} = 2\csc^2\theta.$$

(b)
$$\frac{1}{1-\sin(x)} - \frac{1}{1+\sin(x)} = 2\tan(x)\sec(x).$$

(c)
$$(1+\sec\theta)(1-\cos\theta) = \sin\theta\tan\theta.$$

(d)
$$\frac{1}{\tan\theta + \cot\theta} = \sin\theta\cos\theta.$$

2. Compute the following expressions.

(a) $\cos(75°)$

(b) $\sin(15°)$

6.2 Double-Angle and Half-Angle Formula

Viete (1540-1603) was the first mathematician who derived and used the product-to-sum formula and sum-to-product formula.

Double-Angle and Half-Angle Formula

- $\sin(2x) = 2\sin(x)\cos(x)$
- $\cos(2x) = \cos^2(x) - \sin^2(x) = 2\cos^2(x) - 1 = 1 - 2\sin^2(x)$
- $\cos^2\left(\dfrac{x}{2}\right) = \dfrac{1 + \cos(x)}{2}$
- $\sin^2\left(\dfrac{x}{2}\right) = \dfrac{1 - \cos(x)}{2}$

3. Prove $\tan(2x) = \dfrac{2\tan(x)}{1 - \tan^2(x)}$, using the double-angle formula.

4. If $\tan(\theta) = 3$, evaluate $\tan(3\theta)$.

5. Using $\sin(2x) = 2\sin(x)\cos(x)$, evaluate

$$\sin 6° \cos 12° \cos 24° \cos 48°.$$

6. If $\cos(36°) = \dfrac{a + \sqrt{b}}{c}$ where a, b, and c are relatively prime to each other, find the sum of a, b, and c. (Hint : use the double-angle formula.)

Viete derived the following form by utilizing the sum and difference formula. Product-to-sum formula are known to be used in Calculus, and competition math widely uses this concept, especially in the exams designed for students in grade 10 and above.

$$\sin(u)\sin(v) = \frac{1}{2}\left(\cos(u-v) - \cos(u+v)\right)$$

$$\cos(u)\cos(v) = \frac{1}{2}\left(\cos(u-v) + \cos(u+v)\right)$$

$$\sin(u)\cos(v) = \frac{1}{2}\left(\sin(u+v) + \sin(u-v)\right)$$

$$\cos(u)\sin(v) = \frac{1}{2}\left(\sin(u+v) - \sin(u-v)\right)$$

7. Rewrite the following product into summation of trigonometric expressions.

(a) $\cos(3x)\sin(5x)$

(b) $\sin(5x)\sin(2x)$

8. Evaluate $\cos 10° \sin 20° \sin 40°$. (Hint : use $\sin(u)\cos(v) = \frac{1}{2}(\sin(u+v) + \sin(u-v))$ and $\sin(u)\sin(v) = \frac{1}{2}(\cos(u-v) - \cos(u+v))$.)

On the other hand, we can reverse the sum into the product, which is useful when we deal with trigonometric equations.

$$\sin(\alpha) + \sin(\beta) = 2\sin(\frac{\alpha+\beta}{2})\cos(\frac{\alpha-\beta}{2})$$

$$\sin(\alpha) - \sin(\beta) = 2\cos(\frac{\alpha+\beta}{2})\sin(\frac{\alpha-\beta}{2})$$

$$\cos(\alpha) + \cos(\beta) = 2\cos(\frac{\alpha+\beta}{2})\cos(\frac{\alpha-\beta}{2})$$

$$\cos(\alpha) - \cos(\beta) = -2\sin(\frac{\alpha+\beta}{2})\sin(\frac{\alpha-\beta}{2})$$

9. Find all real solutions of the following equations.

(a) $\sin(3x) + \sin(x) = 0$ for $x \in [0, 2\pi]$ \qquad (b) $\cos(x) + \cos(3x) = 0$ for $x \in [0, 2\pi)$.

10. Turn $\sin(6x) + \sin(4x)$ into the product of sines and cosines.

11. If $\cos(\theta) - \sin(\theta) = \sqrt{2}\sin(20°)$ for acute angle θ, find θ.

12. Evaluate $\dfrac{1 - 4\cos(\dfrac{2\pi}{7})\sin(\dfrac{\pi}{14})}{\sin(\dfrac{\pi}{14})}$. (Hint : the answer is an integer.)

01.

Rewrite $\dfrac{1}{1 + \sin(x)}$ so that it is not in fractional form.

02.

Rewrite $\dfrac{\sin^2(y)}{1 - \cos(y)}$ so that it is not in fractional form.

03.

Use the substitution $x = 2\tan(\theta)$ for $0 < \theta < \dfrac{\pi}{2}$ to write $\sqrt{4 + x^2}$ as a trigonometric function of θ.

04.

Use the substitution $x = 3\cos(\theta)$ for $0 < \theta < \dfrac{\pi}{2}$ to write $\sqrt{9 - x^2}$ as a trigonometric function of θ.

05.

Express $\sin(3x)$ in terms of $\sin(x)$'s.

06.

Express $\cos(3x)$ in terms of $\cos(x)$'s.

07.

Verify the identity $\sec(x) - \cos(x) = \sin(x)\tan(x)$.

08.

Verify the identity $\cos(x) - \dfrac{\cos(x)}{1 - \tan(x)} = \dfrac{\sin(x)\cos(x)}{\sin(x) - \cos(x)}$.

09.

Find the maximum value of $\tan(\alpha - \beta)$ where α, β are acute angles, and $\tan(\alpha) = 4\tan(\beta)$.

10.

Evaluate $\cos^2(55°) + \cos^2(35°)$.

11.

Find the exact value of $\sin(\frac{7\pi}{12})$.

12.

Find the exact value of $\cos(165°)$.

13.

Find the exact value of $\sin(u + v)$ given $\sin(u) = \dfrac{4}{5}$ where $0 < u < \dfrac{\pi}{2}$ and $\cos v = -\dfrac{12}{13}$ where $\dfrac{\pi}{2} < v < \pi$.

14.

Write $\cos(\arctan(1) + \arccos(x))$ as an algebraic expression of x.

15.

Let a and b be angles such that $\sin 2a + \sin 2b = \dfrac{1}{2}$ and $\cos 2a + \cos 2b = \dfrac{\sqrt{3}}{2}$. Evaluate $\cos(2a - 2b)$.

16.

Evaluate $\cos(\frac{\pi}{7}) + \cos(\frac{3\pi}{7}) + \cos(\frac{5\pi}{7})$.

17.

If $\sin 2\theta = \dfrac{16}{25}$ and $\cos\theta > \sin\theta$, then evaluate the value of $\cos\theta - \sin\theta$.

18.

Express $\tan(3x)$ in terms of $\tan(x)$'s.

19.

If $x = \cos(\theta)$ for acute θ satisfies $8x^3 - 6x - 1 = 0$, find the degree measure of θ.

Solution to Core Skill Practice 12

01. $\sec^2(x) - \cot(x)\sec(x)$

02. $1 + \cos(y)$

03. $2\sec(\theta)$

04. $3\sin(\theta)$

05. $\sin(3x) = 3\sin(x) - 4\sin^3(x)$

06. $\cos(3x) = 4\cos^3(x) - 3\cos(x)$

07.

$$
\begin{aligned}
\sec(x) - \cos(x) &= \frac{1}{\cos(x)} - \cos(x) \\
&= \frac{1 - \cos^2(x)}{\cos(x)} \\
&= \frac{\sin^2(x)}{\cos(x)} \\
&= \sin(x)\tan(x)
\end{aligned}
$$

08.

$$
\begin{aligned}
\cos(x) - \frac{\cos(x)}{1 - \tan(x)} &= \frac{\cos(x)(1 - \tan(x)) - \cos(x)}{1 - \tan(x)} \\
&= \frac{\tan(x)\cos(x)}{1 - \tan(x)} \\
&= \frac{\sin(x)\cos(x)}{\sin(x) - \cos(x)}
\end{aligned}
$$

09. $\dfrac{3}{4}$

10. 1

11. $\dfrac{\sqrt{6} + \sqrt{2}}{4}$

12. $-\dfrac{\sqrt{6} + \sqrt{2}}{4}$

13. $-\dfrac{33}{65}$

14. $\dfrac{\sqrt{2}}{2}(x - \sqrt{1 - x^2})$

15. $-\dfrac{1}{2}$

16. $\dfrac{1}{2}$

17. $\dfrac{3}{5}$

18. $\tan(3x) = \dfrac{3\tan(x) - \tan^3(x)}{1 - 3\tan^2(x)}$

19. $\theta = 20°$

TOPIC 7

Application of Trigonometry

It is known that Persian (which is now known as Iran) mathematicians discovered the laws of sines, using half-chords in circles, especially attributed to a famous mathematician called Al-Tusi.

The First Law of Sines (to compute area of triangles)

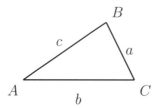

Given the lengths of two sides of a triangle and the measure of the included angle, the area of triangle is defined by

$$\begin{aligned} \text{Area} &= \frac{1}{2}ab\sin(\angle C) \\ &= \frac{1}{2}bc\sin(\angle A) \\ &= \frac{1}{2}ac\sin(\angle B) \end{aligned}$$

The Second Law of Sines (to compute circumradius or connect angle ratio with length ratio)

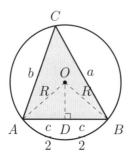

There is an invariant ratio for a triangle such that

$$\frac{a}{\sin A} = \frac{b}{\sin B} = \frac{c}{\sin C} = 2R$$

where the rule is used in the cases for

1. two angles and one side

2. two sides and non-included angle

The law of sine can be easily deduced by using the property of circumcenter, the point of concurrency for perpendicular bisectors.

Al-Kahsi, another Persian mathematician and astronomer, explicitly stated the law of cosines in the 15th century, the form of which corresponds to the modern one of law of cosines. This is the formula we use the most in geometry.

The Law of Cosines (to find angle measures or lengths)

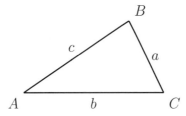

Given the measure of three side lengths, we have the formula

$$a^2 = b^2 + c^2 - 2bc\cos(A)$$
$$b^2 = a^2 + c^2 - 2ac\cos(B)$$
$$c^2 = a^2 + b^2 - 2ab\cos(C)$$

Heron's Formula

Heron's formula is useful to find the area of triangle when the side-lengths are all given, i.e., $A = \sqrt{s(s-a)(s-b)(s-c)}$ where $s = \dfrac{a+b+c}{2}$. Heron's formula, recommended when the side-lengths are integers, can be directly deduced by the law of cosines and the law of sines. The deduction of Heron's formula requires bits of algebra.

$$
\begin{aligned}
\text{Area of } \triangle \text{ ABC} &= \frac{ab}{2}\sin(\theta)\\
&= \frac{1}{2}ab\sqrt{1 - \cos^2(\theta)}\\
&= \frac{1}{2}ab\sqrt{1 - \frac{(c^2 - a^2 - b^2)^2}{4a^2b^2}}\\
&= \sqrt{\frac{4a^2b^2 - (c^2 - a^2 - b^2)^2}{16}}\\
&= \sqrt{\frac{(2ab - c^2 + a^2 + b^2)(2ab + c^2 - a^2 - b^2)}{16}}\\
&= \sqrt{\frac{((a+b)^2 - c^2)(c^2 - (a-b)^2)}{16}}\\
&= \sqrt{\frac{(a+b+c)(-a+b+c)(a-b+c)(a+b-c)}{16}}\\
&= \sqrt{s(s-a)(s-b)(s-c)}
\end{aligned}
$$

1. Find the area of a triangle whose side lengths are 13, 14, and 15.

2. If $\triangle ABC$ has $AB = 5$, $BC = 7$, and $AC = 8$, find the measure of $\angle BAC$ without using a calculator.

3. If $ABCD$ is a square where M and N are the midpoints of $\overline{AB}$ and $\overline{BC}$, respectively. Find $\cos(\angle MDN)$.

4. The diagonals of rectangle $ABCD$ intersect at point P. If $AB = 3$ and $BC = 4$, find $\sin(\angle APB)$.

5. Given an isosceles triangle ABC, where $AC = BC = 7$ and $AB = 2\sqrt{42}$, if D is between A and B, and $CD = 3$, find the length of $\overline{AD}$.

6. If $AB = 4$, $BC = 6$, $CA = 5$ for $\triangle ABC$, find the length of $\overline{AD}$ where D is the midpoint of $\overline{BC}$.

7. Determine whether the triangle ABC that satisfies the following property is right or not.

(a) $a\cos(A) = b\cos(B)$

(b) $\sin(A) + \sin(B) = \sin(C)(\cos(A) + \cos(B))$

8. In triangle ABC, $\angle B = 30°$, $AB = 15$, and $AC = 5\sqrt{3}$. Find the possible degree measures of C.

9. In acute triangle ABC we have $\cos A \cos B = \dfrac{1}{4}$ and $\sin A \sin B = \dfrac{1}{3}$. If the area of triangle ABC is 1, what is the area of the orthic triangle? (The orthic triangle has vertices at the feet of the altitudes of the triangle.) .

10. The perimeter of parallelogram $ABCD$ is 44, and its altitudes are 4 and 7, respectively. Find the possible values of degree measures of A.

Argand and Gauss, who made attempts to consider complex numbers as vectors in the 19th century, built the foundation of two-dimensional vectors. Gauss used this approach to prove the Fundamental Theorem of Algebra, and Hamilton, renowned for Hamiltonian mechanics, suggested that complex numbers must be treated as (a, b), with both direction and magnitude, where (a, b) refers to the horizontal shift of a units to the right and the vertical shift of b units up. He came up with the idea of *quaternions* - a four-dimensional vector that appears in mechanics. It was then further developed by Gibbs, whose notes for his students at Yale University included vector analysis, which now became a modern language of physics and applied mathematics.

Let's have a look of what these great giants said to their pupils and later generations. Vector, denoted by $\vec{v}$ or $\mathbf{v}$, is defined as a quantity that has both a magnitude and a direction. It is usually represented by an arrow in the plane or in space. The length of the arrow is its magnitude and the orientation of the arrow is its direction.

Two arrows with the same magnitude and direction represent the same vector, known as *translates*.

The dot is the initial point and the arrow is the terminal point of the vector. Here, the vector between two points X and Y is denoted $\overrightarrow{XY}$.

X is called the initial point and Y the terminal point of $\overrightarrow{XY}$. The magnitude of the vector $\vec{v}$ is denoted $\|\vec{v}\|$, called either length or norm, which is computed by the usual distance formula. Scaling a vector means changing its length by a scale factor. In fact, this is simply multiplying or dividing a vector quantity by some number. Two non-zero vectors are parallel if a vector is a scalar multiple of the other.

Parallel vectors

$\vec{a} = k\vec{b}$ for some constant k.

Furthermore, if a vector is scaled down by its own length, such vector is called a unit vector, $\hat{v} = \dfrac{\vec{v}}{\|v\|}$, with the length of 1.

Coordinate Vectors

The method of representing a vector in a coordinate plane requires two setups.

1. Tail of the vector at the Origin $(0,0)$

2. Head of the vectors at (a, b)

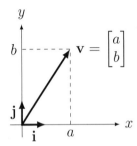

The Cartesian coordinate can be used to represent a vector whose initial point starts at the origin. If the head coordinates are (x, y), then we have two associated vector notations such that

(1) $\begin{bmatrix} x \\ y \end{bmatrix}$ is a column vector.   (2) $\begin{bmatrix} x & y \end{bmatrix}$ is a row vector.

Example

A column vector is useful when a vector is expressed as a linear combination of $\mathbf{i}, \mathbf{j}$, also known as *basis* vectors. For example,

$$\begin{bmatrix} 3 \\ 4 \end{bmatrix} = 3\mathbf{i} + 4\mathbf{j} = 3\begin{bmatrix} 1 \\ 0 \end{bmatrix} + 4\begin{bmatrix} 0 \\ 1 \end{bmatrix} = \begin{bmatrix} 1 & 0 \\ 0 & 1 \end{bmatrix}\begin{bmatrix} 3 \\ 4 \end{bmatrix}$$

---

11. Find the length of the vector $\begin{bmatrix} 4 \\ -3 \end{bmatrix}$.

The length of the vector $\vec{u} - \vec{v}$ or $\vec{v} - \vec{u}$ is found by using the law of cosines. Since $\vec{u}, \vec{v}$ and $\vec{v} - \vec{u}$ form a triangle whose side lengths are $\|\vec{u}\|, \|\vec{v}\|, \|\vec{v} - \vec{u}\|$, we have

$$\|\vec{v} - \vec{u}\|^2 = \|\vec{u}\|^2 + \|\vec{v}\|^2 - 2\|\vec{u}\|\|\vec{v}\|\cos\theta$$

where θ is the angle between $\vec{u}$ and $\vec{v}$. In the figure below, let the horizontal ray be a vector $\vec{u}$ and the one that goes northeast direction be $\vec{v}$. Then, the angle formed by the two rays must be θ. The ray opposite θ must be $\vec{v} - \vec{u}$.

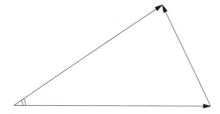

The expression $\|\vec{u}\|\|\vec{v}\|\cos\theta$ is called the dot product of $\|\vec{u}\|$ and $\|\vec{v}\|$, denoted by $\vec{u} \cdot \vec{v}$. We can write it as

$$\|\vec{u} - \vec{v}\|^2 = \|\vec{u}\|^2 + \|\vec{v}\|^2 - 2\vec{u} \cdot \vec{v}$$

Hence, we can write

$$\mathbf{u} \cdot \mathbf{v} = \|\mathbf{u}\|\|\mathbf{v}\|\cos\theta$$

12. Compute the dot product of $\mathbf{u}$ and $\mathbf{v}$ if $\mathbf{u} = \begin{bmatrix} 1 \\ 3 \end{bmatrix}$ and $\mathbf{v} = \begin{bmatrix} 3 \\ -2 \end{bmatrix}$.

13. If $\mathbf{u}$ and $\mathbf{v}$ are unit vectors, and $\mathbf{u} \cdot \mathbf{v} = \dfrac{1}{2}$, find the acute angle between $\mathbf{u}$ and $\mathbf{v}$, in degree measure.

Meaning of $\dfrac{\mathbf{u} \cdot \mathbf{v}}{\|\mathbf{u}\|\|\mathbf{v}\|} = \cos\theta$

Why do we calculate the dot product of two vectors? The dot product is a numerical(scalar) value that is best understood as the amount of projection length of $\mathbf{v}$ onto $\mathbf{u}$. Simply speaking, if we wish to find out "how much" of a vector $\mathbf{v}$ is in a given direction of $\mathbf{u}$, we compute some quantity known as "component" of $\mathbf{v}$ onto $\mathbf{u}$.

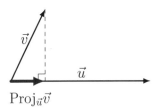

$\text{Proj}_{\vec{u}}\vec{v}$

Let's take a unit vector in a direction of $\mathbf{u}$ and scale it by the scale projection of $\mathbf{v}$ in the $\mathbf{u}$ direction. This is called the projection of $\mathbf{v}$ onto $\mathbf{u}$.

$$\text{Proj}_{\mathbf{u}}\mathbf{v} = \|\mathbf{v}\| \cos\theta \, \frac{\mathbf{u}}{\|\mathbf{u}\|}$$

$$= \|\mathbf{u}\|\|\mathbf{v}\| \cos\theta \, \frac{\mathbf{u}}{\|\mathbf{u}\|^2}$$

$$= \frac{\mathbf{u} \cdot \mathbf{v}}{\|\mathbf{u}\|^2} \mathbf{u}$$

$$= (\mathbf{u} \cdot \mathbf{v}) \frac{\hat{\mathbf{u}}}{\|\mathbf{u}\|}$$

Hence, if $\mathbf{u}$ is a unit vector to begin with, then a dot product of any vector $\mathbf{v}$ to $\mathbf{u}$ gives the amount of projection of $\mathbf{v}$ onto $\mathbf{u}$.

14. Find the projection of $\begin{bmatrix} 3 \\ 7 \end{bmatrix}$ onto $\begin{bmatrix} 2 \\ 1 \end{bmatrix}$.

15. If $\mathbf{a}$ and $\mathbf{b}$ are perpendicular vectors and $\|\mathbf{a}\| = 5$ and $\|\mathbf{b}\| = 3$, find $(2\mathbf{a} - \mathbf{b}) \cdot (4\mathbf{b} + \mathbf{a})$.

16. Find x so that the vectors $\begin{bmatrix} 1 \\ 4 \end{bmatrix}$ and $\begin{bmatrix} x \\ -2 \end{bmatrix}$ are orthogonal.

17. Let $\mathbf{u}$, $\mathbf{v}$, and $\mathbf{w}$ be vectors such that $\|\mathbf{u}\| = 1$, $\|\mathbf{v}\| = 2$, and $\|\mathbf{w}\| = 3$, and

$$\mathbf{u} + \mathbf{v} + \mathbf{w} = \mathbf{0}.$$

Compute $\mathbf{u} \cdot \mathbf{v} + \mathbf{u} \cdot \mathbf{w} + \mathbf{v} \cdot \mathbf{w}$.

We normally refer a position vector to a vector that originates from the origin O. Then, we can deduce the vectors connecting the three points O, A, B as in the following diagram.

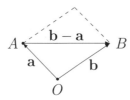

Notice that we do not break away from the usual vector algebra, which uses parallelogram law or triangle law. In fact, $\mathbf{a} + (\mathbf{b} - \mathbf{a}) = \mathbf{b}$.

Golden Rule

$$\overrightarrow{OA} + \overrightarrow{AB} = \overrightarrow{OB} \qquad\qquad \overrightarrow{OB} - \overrightarrow{OA} = \overrightarrow{AB}$$

18. In the following diagram, the position vectors of points A and B relative to O are $\mathbf{a}$ and $\mathbf{b}$, respectively. Assume that the lines AB and OP intersect at Q. Given that $\overrightarrow{BP} = 3(\mathbf{a} + \mathbf{b})$ and $\overrightarrow{AQ} = r\overrightarrow{AB}$ and $\overrightarrow{OQ} = s\overrightarrow{OP}$, evaluate r and s.

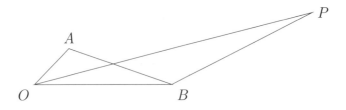

19. The points A, B, and C have position vectors $\mathbf{u} + \mathbf{v}$, $3\mathbf{u} - 2\mathbf{v}$, and $6\mathbf{u} + m\mathbf{v}$ relative to the origin O. Find the value of m for which A, B, and C are collinear.

20. The points A and B have position vectors $4\mathbf{i} + 3\mathbf{j}$ and $\mathbf{i} + t\mathbf{j}$, respectively. If $\cos \angle AOB = \dfrac{2}{\sqrt{5}}$, find the values of t.

21. $|a\mathbf{i} + b\mathbf{j}| = 5$ and $a\mathbf{i} + b\mathbf{j}$ is perpendicular to $8\mathbf{i} - 6\mathbf{j}$. Find the value of $|a|$ and of $|b|$.

22. Given an equilateral ABC with the side length of 2, if there is a circle centered at A that passes through both B and C, and $\overrightarrow{AP} \cdot \overrightarrow{BC} = -4$ for a point P on the circle, then compute PC.

23. Given points $A(3,0)$ and $B(3,4)$, if P is a point on the circle $x^2 + y^2 = 1$, find the minimum value of $|\overrightarrow{PA} + \overrightarrow{PB}|$.

24. Let O and H denote the circumcenter and orthocenter of triangle ABC, respectively. If we let O as the origin, then it is well-known that $\mathbf{OH} = \mathbf{OA} + \mathbf{OB} + \mathbf{OC}$. If $AO = AH$, find the measure of acute $\angle A$.

Complex numbers appear most in quadratic equations, but they were born while studying cubic equations. It all began with Cardano, who came up with a general formula for cubic equations $x^3 = px + q$ with some imaginary numbers appearing inside radicals. When Cardano came up with a method of solving these cubic equations, he found a bizarre case of having negatives inside the radical expression, which he called it as *casus irreducibilis*. Now, the Europeans saw the existence of negatives in the radicals. Descartes also noticed the existence of imaginary number. John Wallis, who came up with the modern notion of a real number line with 0, negatives, and positives, advanced the geometric interpretation of $\sqrt{-1}$. All of these attempts saw their hiatus when De Moivre and Newton used De Moivre's theorem to solve Cardano's irreducible cases, and Euler introduced the notion of $i = \sqrt{-1}$. In modern days, we see i appearing in quadratic equations. If we solve the quadratic equation $x^2 - 2x + 3 = 0$, for instance, then we get using the quadratic formula

$$x = \frac{2 \pm \sqrt{-8}}{2}$$

Here, $\sqrt{-8}$ cannot be found normally in a real line, but if we define $i = \sqrt{-1}$, then $\sqrt{-8} = \sqrt{-1 \times 8} = \sqrt{8i^2} = \sqrt{8}\sqrt{i^2} = 2\sqrt{2}i$. Hence, we can conclude

$$x = \frac{2 \pm 2\sqrt{2}i}{2}$$
$$= 1 \pm \sqrt{2}i$$

This is a good example of a complex number z. The complex number $1 + \sqrt{2}i$ has real part 1 and imaginary part $\sqrt{2}$. The set of real numbers, denoted by $\mathbb{R}$, is a subset of the set of complex numbers since every real number can be written as $a + 0i$. On the other hand, if the real part is 0, then z is called *purely imaginary*.

If $z = a + bi$, then the real part of z is a and the imaginary part of z is b, i.e., $z = -3 + 7i$ with Re(z)=-3 and Im(z)=7.

25. Let a and b be nonzero real numbers such that

$$(3 - 5i)(a + bi)$$

is pure imaginary. Find $\dfrac{a}{b}$.

What distinguishes complex numbers from real numbers? Complex numbers themselves cannot be compared. Hence, we need to introduce the notion of magnitude. Here is the definition of the length of a complex number $z = a + bi$, i.e.

$$|z|^2 = a^2 + b^2 = (a + bi)(a - bi)$$

where $|z|$ or $||z||$ refer to the magnitude of z. As one can see from the expression above, we have $a - bi$. If $z = a + bi$, then the complex conjugate is defined as $\overline{z} = a - bi$. In school Precalculus, we only use conjugates when complex numbers are divided, i.e., if we divide $\dfrac{3 + 2i}{4 - i}$, then we multiply both the top and the bottom by the complex conjugate of the denominator.

Furthermore, conjugates can be used in factoring quadratic equations. For instance, $x^2 + 4 = (x - 2i)(x + 2i)$; $2x^2 + 50 = 2(x^2 + 25) = 2(x + 5i)(x - 5i)$; $x^2 + 1 = (x - i)(x + i)$. Also, if a complex number is multiplied by its conjugate, then the result is always a real number. The following list shows the powers of i, which is cyclic of four.

Powers of i

1. $i = \sqrt{-1}$

2. $i^2 = -1$

3. $i^3 = -i$

4. $i^4 = 1$

We find a pattern $i, -1, -i, 1, i, -1, -i, 1, \cdots$. Therefore, we can easily solve $i^{19} = i^{16} \times i^3 = 1 \times i^3 = -i$. Similarly, we can find $i^{4n+3} = i^{4n} \times i^3 = (i^4)^n \times -i = 1^n \times -i = -i$.

26. Compute $i^1 + i^2 + i^3 + \cdots + i^{2020} + i^{2021}$.

27. Let z be a complex number such that

$$|z|^2 = 2 + 2i - z^2.$$

Find $|z|^2$.

28. Let u and v be nonzero complex numbers such that

$$|u| = |v| = |u + v|.$$

Find the sum $\dfrac{u}{v} + \dfrac{v}{u}$.

Carl Friedrich Gauss, also known as *princeps mathematicorum*, formulated the **Fundamental Theorem of Algebra**, though difficult to prove in highschool mathematics, ensuring that we may have n number of complex roots for nth degree polynomial.

F.T.A

A polynomial equation $p(x) = 0$ of degree n has exactly n complex roots, some of which may be repeated roots.

If all of the coefficients are real, then complex roots will always occur in conjugate pairs. Similarly, if all of the coefficients are complex, we have at least one complex root. This is an important notion because it guarantees that we will not get any number other than complex number. If we start from rational coefficients, we may end up getting irrational root. Consider $x^2 - 5 = 0$. Also, we can even get a complex root if we start with rational coefficients, i.e., $x^2 + 1 = 0$. On the hand, if we let complex coefficients, then we end up getting complex roots whatsoever. In this sense, we say $\mathbb{C}$ is closed.

A short yet powerful application of the theorem has to deal with the existence of a polynomial function. For an instance, think about $f(x) \in \mathbb{Z}[x]^1$ where $f(x) = a_2 x^2 + a_1 x + a_0$. Assume $f(r) = f(s) = f(t) = 0$ for some distinct complex constants, r, s, and t. Then, the theorem ensures that $f(x)$ cannot be quadratic. In fact, $f(x) = 0$ for all $x \in \mathbb{C}$. The reason why this is true is because $f(x) = (x - \alpha)(x - \beta)$ for some complex α and β, allowing the case of $\alpha = \beta$, according to the theorem. Hence, at least two of r, s, and t must be repeated, but it contradicts the assumption that all zeros are distinct. This type of question usually comes out in math competitions, especially asking about the existence of polynomial functions.

29. If $1 - 2i$ is one root of the quadratic equation $x^2 - 2x + 5 = 0$, then find the other root.

---

[1]This means that $f(x)$ is the polynomial whose coefficients are integers, where the set of integers is denoted by $\mathbb{Z}$.

30. Show, using the factor theorem, that the linear expressions below are factors of the given polynomial, and determine the other factor by inspection.

$$x - i, \ x + 2i \text{ are factors of } 4x^3 + ix^2 + 11x - 6i.$$

31. Let z be a complex number satisfying $z^2 = 6z + 6 - 8i$. Find z where $|z|$ is largest.

A complex number can be represented on an Argand diagram. Argand, born in Switzerland, was a bookkeeper at Paris, still unknown whether he had proper mathematical training. However, he produced a pamphlet without his name, dealing with the famous Argand diagram, and one of the copies ended up in the hands of Legendre, who then forwarded it to his colleagues. Luckily, mathematics society figured out who came up with the diagram, and now we will learn about his idea laid out on his pamphlet with the title of *Essay on the Geometrical Interpretation of Imaginary Quantities*. Now, in the following diagram, the x-axis represents the real part and the y-axis the imaginary part.

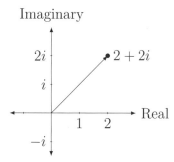

Apart from rectangular(or Cartesian) form, we can write complex numbers in polar form using trigonometry. Thanks to Euler, he visualized complex numbers as dots with rectangular coordinates. For the complex number $z = a + bi$,

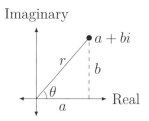

where $a = r\cos\theta$ and $b = r\sin\theta$. Therefore,
$a + bi = r\cos\theta + ir\sin\theta = r(\cos\theta + i\sin\theta) = r\text{cis}\theta$. Here, "cis" is abbreviated for $\cos + i\sin$.

The modulus of z is r, the distance from the origin, written as $|z|$. The argument of z is θ, the angle between the real axis and a line joining z to the origin, written as $\arg(z) = \theta$.

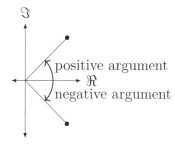

This way of writing a complex number is often called the modulus-argument form. Also, notice from the figure that the argument is best understood at $-180° < \theta \le 180°$ or $-\pi < \theta \le \pi$.

32. Convert the following complex numbers and write them in rectangular form.

(a) $2\text{cis}(45°)$

(b) $20\text{cis}(-135°)$

(c) $5\text{cis}\left(-\dfrac{\pi}{3}\right)$

(d) $3\text{cis}(\pi)$

Hamilton, the same physicist and mathematician, who used the notion of vector, wrote on his memoir, defining ordered pairs of real numbers (a, b) for $a + bi$ to be a *couple*. He defined addition and multiplication of couples such that

- $(a, b) + (c, d) = (a + c, b + d)$

- $(a, b) \cdot (c, d) = (ac - bd, bc + ad)$

Guess what! His *couple* describes the definition of complex numbers. Then, who came up with this geometric interpretation of complex numbers, especially tying up angles and magnitudes? His name is Caspar Wessel, a Norwegian mathematician whose work went unnoticed even after his death. He used the parallelogram law and came up with the idea of adding polar angles and multiplying the magnitudes. Let's have a look. If $z_1 = r_1 \text{cis}(\theta_1)$ and $z_2 = r_2 \text{cis}(\theta_2)$, then

1. $z_1 z_2 = r_1 r_2 \text{cis}(\theta_1 + \theta_2)$

2. $\dfrac{z_1}{z_2} = \dfrac{r_1}{r_2} \text{cis}(\theta_1 - \theta_2)$

or, equivalently we may have for $z_1 = [r_1, \theta_1]$ and $z_2 = [r_2, \theta_2]$,

1. $z_1 z_2 = [r_1 r_2, \theta_1 + \theta_2]$

2. $\dfrac{z_1}{z_2} = [\dfrac{r_1}{r_2}, \theta_1 - \theta_2]$

Note that if the argument is not in the range $-\pi < \theta \le \pi$, then we add or subtract the multiples of 2π until it is in this range.

33. Evaluate the following expressions.

(a) $2\text{cis}(30°) \times 3\text{cis}(60°)$

(b) $\dfrac{4\text{cis}(120°)}{2\text{cis}(30°)}$

A close friend of Newton, De Moivre contributed much to probability and trigonometry. He came up with De Moivre's Theorem and this result was generalized by Euler later on. This is usually used to raise complex numbers to integral powers, i.e., $(2 + 3i)^4$. The reason why we learn about complex numbers in polar form is because the multiplication of complex numbers is simply rotation by adding or subtracting the given arguments, reducing large numbers into something that we know, thanks to periodicity of angular rotation.

Why is this so useful? Think about it this way. If we have extremely large real numbers, it is difficult for us to comprehend what such numbers could do. Hence, we wrap the set of real numbers around the unit circle to understand real number behavior, now having periodicity. This was how trigonometry showed its significance - producing periodicity to a set of real numbers. This theorem bridges between trigonometry and complex numbers, which had different beginnings.

$$(\cos\theta + i\sin\theta)^n = \cos(n\theta) + i\sin(n\theta)$$
$$(r\text{cis}(\theta))^n = r^n\text{cis}(n\theta)$$

34. Simplify $(2\text{cis}(45°))^5$.

35. Use De Moivre's Theorem to evaluate

(a) $(1 + i)^{12}$

(b) $(1 - i)^{24}$

Applying De Moivre's Theorem, we can solve the equation of the form $z^n = a + bi$ where z and $a + bi$ are complex numbers and n is any integer. When we solve equations of this form, the following rule is helpful.

1. Write the rectangular form $a + bi$ into a polar form $r \text{cis} \theta$, where $r = |z| = \sqrt{a^2 + b^2}$.

2. Write a general form of $r \text{cis} \theta$, i.e., $z^n = r \text{cis}(\theta + 2\pi k)$.

3. Apply De Moivre's Theorem so that $z = r^{\frac{1}{n}} \text{cis}(\dfrac{\theta + 2\pi k}{n})$.

4. Substitute values of $k = 0, 1, \cdots, n-1$ to give all values of z (hopefully, in the range of $-\pi < \theta \leq \pi$).

Notice that by the Fundamental Theorem of Algebra, we expect to find n complex roots for $z^n = a + bi$.

36. Solve the following equations for complex z.

(a) $z^3 - 1 = 0$

(b) $z^4 - 1 = 0$

(c) $z^6 - 1 = 0$

(d) $z^3 - 8 = 0$

Euler, using De Moivre's theorem, visualized the roots of $z^n = 1$ as vertices of regular n-gon. He went a bit further than this, defining exponential form of complex number, i.e,
$e^{i\theta} = \cos(\theta) + i\sin(\theta)$, which we now know as Euler's Formula. The famous Euler's identity uses π, i, e, 1, and 0.

$$e^{\pi i} + 1 = 0.$$

- $\cos(n\theta) + i\sin(n\theta) = e^{i(n\theta)} = (e^{i\theta})^n = (\cos(\theta) + i\sin(\theta))^n$

- $\cos(\theta) = \dfrac{e^{i\theta} + e^{-i\theta}}{2}$

- $\sin(\theta) = \dfrac{e^{i\theta} - e^{-i\theta}}{2i}$

This conversion is extremely useful in solving series of trigonometric expressions, and as anyone can expect, it oftentimes appears in competition math.

37. Evaluate
$$\sum_{n=0}^{\infty} \frac{\cos n\theta}{2^n},$$
where $\cos\theta = \dfrac{2}{7}$.

38. If $(\cos(x))^4 = a\cos(4x) + b\cos(2x) + c$, find $a + b + c$.

Core Skill Practice 13

01.

Solve the cubic equation $x^3 + x^2 - 2 = 0$.

02.

A quadratic equation has roots $z_1 = 3 + 3i$ and $z_2 = 3 - 3i$. Find the quadratic equation with the leading coefficient of 1.

03.

Find the quartic equation with integer coefficients whose roots are $i, -i$ and 1(repeated), whose leading coefficient is 1.

04.

Determine the other factor by inspection.

$$x - i, x + 2i \text{ are factors of } x^3 + 2ix^2 + x + 2i.$$

05.

Write down a polynomial with leading coefficient of 1 with roots 2, $3i$, over $\mathbb{C}$ and over $\mathbb{R}$, respectively.

06.

Given that $z = 2 + 3i$ is a solution of $z^4 - 5z^3 + 18z^2 - 17z + 13 = 0$, determine the other three solutions.

07.

Find the length of $z = -1 + 2i$.

08.

Write the complex number $z = 2 + 2\sqrt{3}i$ in polar form.

09.

Write the complex number in rectangular form $a + bi$

$$z = \sqrt{8}\left[\cos\left(-\frac{\pi}{6}\right) + i\sin\left(-\frac{\pi}{6}\right)\right]$$

10.

Find the product $z_1 z_2$ of the complex numbers where $z_1 = 2(\cos \frac{\pi}{3} + i \sin \frac{\pi}{3})$ and $z_2 = 4(\cos \frac{4\pi}{3} + i \sin \frac{4\pi}{3})$.

11.

Find the quotient $\frac{z_1}{z_2}$ of the complex numbers where $z_1 = 8(\cos(240°) + i \sin(240°))$ and $z_2 = 4(\cos(90°) + i \sin(90°))$.

12.

Use De Moivre's Theorem to find $(1 - \sqrt{3}i)^{10}$.

13.

Find z^{10} of the complex number $z = 1 + \sqrt{3}i$.

14.

Find z^3 of the complex number $z = \sqrt{2}(\cos(120°) + i\sin(120°))$.

15.

If $z^3 = 2(\cos\dfrac{2\pi}{3} + i\sin\dfrac{2\pi}{3})$, then find z.

16.

Find the complex number that rotates a complex number "$z = \sqrt{3} + i$" 30° about the Origin.

17.

Compute $e^{2\pi i/3}$.

18.

If $e^{i\alpha} = \dfrac{4}{5} + \dfrac{3}{5}i$ and $e^{i\beta} = \dfrac{5}{13} + \dfrac{12}{13}i$, then find $\cos(\alpha + \beta)$.

19.

The set of two-dimensional vectors $\mathbf{x}$ such that

$$\mathbf{x} \cdot \mathbf{x} = \mathbf{x} \cdot \begin{pmatrix} 0 \\ 4 \end{pmatrix}$$

form a closed figure in the plane. Find the area of the region contained in the figure.

20.

The complex numbers u and v satisfy

$$u\bar{v} = 3 + 4i.$$

Find $\bar{u}v$.

01. $x = 1, -1 + i, -1 - i$

02. $x^2 - 6x + 18 = 0$

03. $x^4 - 2x^3 + 2x^2 - 2x + 1 = 0$

04. $x + i$

05. $(x - 2)(x - 3i) = x^2 - (2 + 3i)x + 6i$ over $\mathbb{C}$ and $(x - 2)(x - 3i)(x + 3i) = x^3 - 2x^2 + 9x - 18$ over $\mathbb{R}$.

06. $z = 2 - 3i, \dfrac{1 + \sqrt{3}i}{2}, \dfrac{1 - \sqrt{3}i}{2}$

07. $|z| = \sqrt{(-1)^2 + 2^2} = \sqrt{5}$

08. $z = 4\mathrm{cis}\left(\dfrac{\pi}{3}\right)$

09. $z = \sqrt{6} - \sqrt{2}i$

10. $z_1 z_2 = 4 - 4\sqrt{3}i$

11. $\dfrac{z_1}{z_2} = -\sqrt{3} + i$

12. $-512 + 512\sqrt{3}i$

13. $z^{10} = -512 - 512\sqrt{3}i$

14. $z^3 = 2\sqrt{2}$

15. $z = \sqrt[3]{2}\mathrm{cis}\left(\dfrac{2\pi}{9}\right), \sqrt[3]{2}\mathrm{cis}\left(\dfrac{8\pi}{9}\right), \sqrt[3]{2}\mathrm{cis}\left(\dfrac{14\pi}{9}\right)$

16. $1 + \sqrt{3}i$

17. $-\dfrac{1}{2} + \dfrac{\sqrt{3}}{2}i$

18. $\cos(\alpha + \beta) = -\dfrac{16}{65}$

19. Let $\mathbf{x} = <a, b>$. Then, $a^2 + b^2 = 4b$. Hence, $a^2 + b^2 - 4b + 4 = 4$. The area of a circle bounded by this region is 4π.

20. $\overline{u\overline{v}} = \overline{3 + 4i} = 3 - 4i = \overline{u}v$

TOPIC
8

Polar Coordinates and Graphs

Cavalieri and Pascal started to use the concept of polar coordinates in 1600's, and Bernoulli is mainly responsible for the invention of polar coordinates in the late 17th century. In the early development of Calculus, rectangular coordinates were used instead of polar coordinates. Nevertheless, the simplification of expressions when circular or cylindrical symmetry was involved spurred the development of polar expressions. When we plot points with the Cartesian coordinates, we typically identify the location of each point with coordinates that denote where the point is horizontally and vertically with respect to a point we call the origin. These are the x- and y-coordinates that you are already used to; we sometimes refer to these as rectangular coordinates. While this is the most common method we use to identify the location of a point on a plane, it is not the only method. As any reader could guess, the founder of Calculus, Issac Newton, must have known about the polar system, examining the transformation between the two coordinate systems.

Polar coordinates offer another method for denoting the location of a point on a plane. As with rectangular coordinates, we start with a point we call the origin and we identify the location of each point with an ordered pair, (r, θ). The r-coordinate is the distance from the point to the origin; we call this coordinate the radial coordinate. The θ-coordinate is the angular coordinate, which we define in the same way we related angles to points on the unit circle.

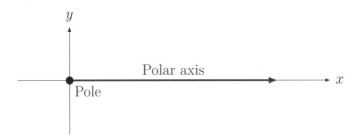

In a polar coordinate system, choose a fixed point, called the pole, and a ray from the pole called the polar axis. An ordered pair is the polar coordinates of a point.

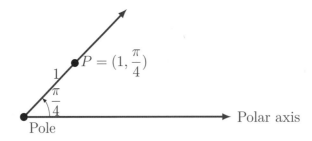

In order to locate a polar point,

    1. Rotate θ radian **counterclockwise**.

    2. Reach out for a distance of r units along the terminal side of the angle (in radians).

A point (r, θ) does not have a unique representation. In fact, for the same point, there are infinitely many representations such as $(r, \theta \pm 2\pi n)$ or $(-r, \theta \pm 2\pi n + \pi)$.

1. Plot the following polar coordinates.

(a) $\left(2, \dfrac{\pi}{3}\right)$

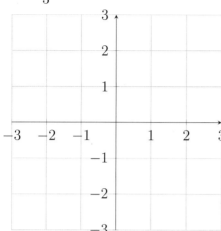

(b) $\left(-2, \dfrac{\pi}{3}\right)$

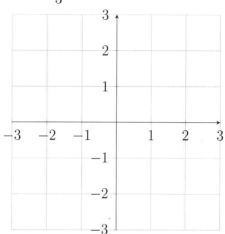

(c) $\left(2, -\dfrac{\pi}{3}\right)$

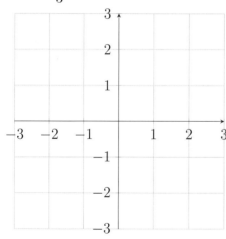

(d) $\left(-2, -\dfrac{\pi}{3}\right)$

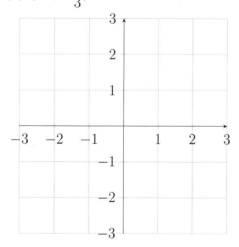

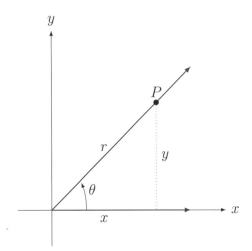

As the figure shows, polar (r, θ) is directly connected to rectangular (x, y) by a right triangle. Let's deduce the formula in order to convert from one form to the other.

Polar $\Longrightarrow$ Rectangular

If a point P with polar coordinates (r, θ), the rectangular coordinates of P satisfy

$$x = r\cos(\theta)$$
$$y = r\sin(\theta)$$

2. Convert the following polar coordinates into rectangular coordinates.

(a) $(3, \dfrac{\pi}{4})$ (b) $(-2, -\dfrac{2\pi}{3})$

Rectangular $\Longrightarrow$ Polar

Given rectangular coordinates (x, y), the polar coordinates of P can be found by

$$r = \sqrt{x^2 + y^2}$$
$$\tan(\theta) = \frac{y}{x} \text{ where } -\frac{\pi}{2} < \theta < \frac{\pi}{2}$$

This method works for points P off the two axes. In order to be precise, follow the following steps.

1. Plot (x, y) and see if the point is on or off the two coordinate axes.

2. If $x = 0$ or $y = 0$, find r directly with the nonzero coordinate value. If not, use $r = \sqrt{x^2 + y^2}$.

3. If $x = 0$ or $y = 0$, find θ directly, according to r's sign. Otherwise, analyze the quadrant where the point lies in.

If the point is in either 1st or 4th quadrant, then use $\theta = \arctan(\frac{y}{x})$. Otherwise, use $\theta = \pi + \arctan(\frac{y}{x})$.

3. Let r be positive. Convert the following rectangular coordinates into polar coordinates.

(a) $(\sqrt{3}, -1)$

(b) $(2\sqrt{2}, -2\sqrt{2})$

4. Convert the following polar coordinates into rectangular coordinates.

(a) $\left(4, \dfrac{\pi}{3}\right)$

(b) $\left(-2, -\dfrac{\pi}{3}\right)$

5. Find other polar coordinates for the given polar coordinate $\left(2, \dfrac{\pi}{3}\right)$ such that

(a) $r > 0$ where $2\pi \le \theta < 3\pi$.

(b) $r < 0$ where $0 \le \theta < 2\pi$.

8.3 Transforming Equations

The following two methods will transform most polar forms into rectangular forms. From rectangular to polar, simply substitute $x = r\cos\theta$ and $y = r\sin\theta$.

1. Multiplying r to both sides of the equation

2. Squaring both sides of the equation.

Example

Change $x^2 + y^2 = 4$ into polar form.

Solution
Since $x = r\cos(\theta)$ and $y = r\sin(\theta)$, we substitute the two expressions into the given equation.
$$r^2\cos^2(\theta) + r^2\sin^2(\theta) = r^2 = 4$$
Hence, we get $r = 2$. Normally, we change the equation into a circle.

6.

(a) Change $x + y = 3$ into polar form.

(b) Convert $xy = 4$ into polar form.

7.

(a) Change $r = 3\cos(\theta)$ into rectangular form.

(b) Convert $r = 2$ into rectangular form.

In a polar plane, the line that passes through the pole whose slope is m is the set of points satisfying $\tan\theta_0 = m$ where θ_0 is the argument. Hence, the polar equation of a line is $\theta = \theta_0$. For instance, $\theta = \dfrac{\pi}{6}$ is a line $y = \dfrac{1}{\sqrt{3}}x$.

On the other hand, given a line l, a locus of point $P(r,\theta)$ that does not pass through the Origin yet with the perpendicular foot $N(r_0,\theta_0)$ from the Origin to the line, we have

$$\overline{OP} \cdot \cos(\angle NOP) = \overline{ON}$$

- If $r > 0, r_0 > 0$, then $r\cos(\theta - \theta_0) = r_0$.

- If $r > 0, r_0 < 0$, then $(r_0, \theta_0) = (-r_0, \theta_0 + \pi)$ implies $r\cos(\theta - \theta_0 - \pi) = -r_0$. Hence, we also get $r\cos(\theta - \theta_0) = r_0$.

- If $r < 0, r_0 > 0$ and $r < 0, r_0 < 0$, then $r\cos(\theta - \theta_0) = r_0$.

If $r_0 \neq 0$, the line passing through $N(r_0, \theta_0)$ that is perpendicular to $\overline{ON}$ has the polar equation of a line

$$r\cos(\theta - \theta_0) = r_0 \text{ or } y = -\cot\theta_0 x + r_0\csc\theta_0$$

8. Find the Cartesian equation for $r\sin(\theta + \dfrac{\pi}{6}) = 3$.

At a Cartesian plane, we know that three points are collinear if the slopes between any two are equal. Back to our discussion, at a polar plane, what is the condition necessary for $P(r_1, \theta_1), Q(r_2, \theta_2), R(r_3, \theta_3)$ to be collinear? We will bring the notion of vector and complex numbers into play! Yes, vector and complex numbers do return to this subject.

The necessary-sufficient condition for $P(r_1, \theta_1), Q(r_2, \theta_2), R(r_3, \theta_3)$ is

$$r_3 e^{i\theta_3} - r_1 e^{i\theta_1} = t(r_3 e^{i\theta_3} - r_2 e^{i\theta_2})(\text{ or } \mathbf{v_1} = k\mathbf{v_2})$$

for some real number t where the complex number that corresponds to P, Q, R is $r_1 e^{i\theta_1}, r_2 e^{i\theta_2}, r_3 e^{i\theta_3}$, respectively. Guess why this is true. A vector $\overrightarrow{PR}$ is parallel to $\overrightarrow{QR}$. Now, let's look at closely what this means. It means that

$$\frac{r_3 e^{i\theta_3} - r_1 e^{i\theta_1}}{r_3 e^{i\theta_3} - r_2 e^{i\theta_2}}$$

is a real number. In other words,

$$\text{Im}\left(\frac{r_3 e^{i\theta_3} - r_1 e^{i\theta_1}}{r_3 e^{i\theta_3} - r_2 e^{i\theta_2}}\right) = 0 \implies \text{Im}((r_3 e^{i\theta_3} - r_1 e^{i\theta_1})(r_3 e^{-i\theta_3} - r_2 e^{-i\theta_2})) = 0$$

$$\implies -r_3 r_2 \sin(\theta_3 - \theta_2) - r_1 r_3 \sin(\theta_1 - \theta_3) + r_1 r_2 \sin(\theta_1 - \theta_2) = 0$$

$$\implies -r_3 r_2 \sin(\theta_3 - \theta_2) - r_1 r_3 \sin(\theta_1 - \theta_3) - r_1 r_2 \sin(\theta_2 - \theta_1) = 0$$

$$\implies r_3 r_2 \sin(\theta_3 - \theta_2) + r_1 r_3 \sin(\theta_1 - \theta_3) + r_1 r_2 \sin(\theta_2 - \theta_1) = 0$$

$$\implies r_1 r_3 \sin(\theta_1 - \theta_3) + r_3 r_2 \sin(\theta_3 - \theta_2) + r_2 r_1 \sin(\theta_2 - \theta_1) = 0$$

You will wonder how the first implication works. Have a look at the following complex arithmetic.

$$\frac{z}{w} = \frac{z \cdot \overline{w}}{w \cdot \overline{w}} = \frac{z \cdot \overline{w}}{|w|^2}$$

All we care about is the imaginary part, so the denominator (real number) does not play a significant role in computation. Hence, the necessary-sufficient condition is

$$r_1 r_3 \sin(\theta_1 - \theta_3) + r_3 r_2 \sin(\theta_3 - \theta_2) + r_2 r_1 \sin(\theta_2 - \theta_1) = 0$$

Therefore, if we have to find the polar equation that passes through two distinct points $P(r_1, \theta_1), Q(r_2 \theta_2)$, then (r, θ) satisfies

$$r_1 r \sin(\theta_1 - \theta) + r r_2 \sin(\theta - \theta_2) + r_2 r_1 \sin(\theta_2 - \theta_1) = 0$$

Three points P, Q, and R are collinear if and only if

$$r_1 r_3 \sin(\theta_1 - \theta_3) + r_3 r_2 \sin(\theta_3 - \theta_2) + r_2 r_1 \sin(\theta_2 - \theta_1) = 0$$

The polar equation with (r, θ) that passes through two distinct points $P(r_1, \theta_1), Q(r_2 \theta_2)$ is equal to

$$r_1 r \sin(\theta_1 - \theta) + r r_2 \sin(\theta - \theta_2) + r_2 r_1 \sin(\theta_2 - \theta_1) = 0$$

9. Determine whether the three points are collinear.

$$\left(1, \frac{\pi}{4}\right), \left(2, \frac{5\pi}{4}\right), \left(-1, \frac{\pi}{4}\right)$$

10. Find a polar equation that passes through the following two points.

$$\left(1, \frac{\pi}{4}\right), \left(2, \frac{\pi}{3}\right)$$

We cannot help but use polar form of complex numbers into the study of polar equations. Let's have a look at circle, the simplest form of conic sections. If r is fixed, then it must form a circle. Hence, the circle equation whose radius is K, centered at the pole is

$$r = K \text{ or } r = -K$$

Before generalizing the circle equation, let's find out the distance between two distinct points $P(r_1, \theta_1), Q(r_2, \theta_2)$. Assume that the complex number associated[1] to P and Q is $r_1 e^{i\theta_1}$, $r_2 e^{i\theta_2}$, respectively. Recall that, given a complex number z, we get

$$|z|^2 = z \cdot \overline{z}$$

Now, we let $PQ = z = r_1 e^{i\theta_1} - r_2 e^{i\theta_2}$, then $\overline{z} = r_1 e^{-i\theta_1} - r_2 e^{-i\theta_2}$. Hence,

$$
\begin{aligned}
PQ^2 &= |r_1 e^{i\theta_1} - r_2 e^{i\theta_2}|^2 \\
&= (r_1 e^{i\theta_1} - r_2 e^{i\theta_2})(r_1 e^{-i\theta_1} - r_2 e^{-i\theta_2}) \\
&= r_1^2 + r_2^2 - (r_1 r_2 e^{i(\theta_1 - \theta_2)} + r_1 r_2 e^{-i(\theta_1 - \theta_2)}) \\
&= r_1^2 + r_2^2 - r_1 r_2 (e^{i(\theta_1 - \theta_2)} + e^{-i(\theta_1 - \theta_2)}) \\
&= r_1^2 + r_2^2 - 2 r_1 r_2 \cos(\theta_1 - \theta_2)
\end{aligned}
$$

11. Find the square of the distance between two polar points $\left(3, \dfrac{\pi}{5}\right)$ and $\left(7, \dfrac{8\pi}{15}\right)$.

---

[1]Get used to how we use the tool of complex number expressions to polar forms. It is quite handy ever since Euler brought it into our toolbox.

If the center is given by (r_0, θ_0) at a polar plane, then the circle equation whose radius is a is equal to

$$\sqrt{r^2 + r_0^2 - 2rr_0 \cos(\theta - \theta_0)} = a$$

Hence, the general equation can be written by

$$r^2 + r_0^2 - 2rr_0 \cos(\theta - \theta_0) = a^2$$

12. Find a polar equation of the circle whose radius is 3 and the center is $\left(3, \dfrac{\pi}{4}\right)$.

One of the great benefits we take from learning polar expressions is that conics can be simply expressed using polar expressions. Let's have a look at the following diagram.

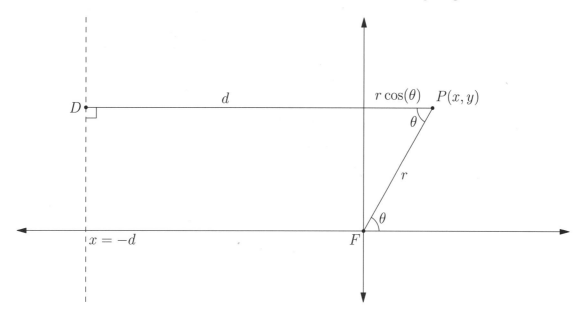

In conic section, *eccentricity* **e** is defined as the ratio of PF to PD where P is a point on conic curve, F is a focus(a fixed point), and D is a point on the directrix(a given line). As one can see from the figure above, we can deduce that

- $PF = r = \sqrt{x^2 + y^2}$

- $PD = r\cos(\theta) + d$

Since $e = \dfrac{PF}{PD}$, we use algebra to get the general formula for polar expressions of conic sections, i.e.

$$e = \frac{PF}{PD}$$
$$ePD = PF$$
$$er\cos(\theta) + ed = r$$
$$ed = r(1 - e\cos(\theta)$$
$$\frac{ed}{1 - e\cos(\theta)} = r$$

Hence, we get the general formula $r = \dfrac{ed}{1 - e\cos(\theta)} = \dfrac{K}{1 - e\cos(\theta)}$, where $K = ed$. Since e and d are fixed numbers, we can consider K as a constant.

We will investigate this expression so that we can understand the shape of conic sections.

- if $e = 0$, then $r = K$ implies that we have a circle.

- if $0 < e < 1$, then $1 - e\cos(\theta) \neq 0$ for any θ, so we can draw a curve when $\theta = 0$. The maximum value of r occurs when the denominator is the smallest, especially when $\cos(\theta) = 1$. This implies that we get an ellipse, the only closed curve out of hyperbola, parabola, and ellipse.

- if $e = 1 = PF/PD$, this fits the definition of parabola. Also, $\cos(\theta) \neq 1$ means $\theta \neq 0$.

- if $e > 1$, then $\cos(\theta) \neq 1/e$. If $\cos(\theta) < 1/e$, then $1 - e\cos(\theta) > 0$ and $\cos(\theta) \neq 1/e$ implies that the graph of the curve cannot get inside certain direction. If we paraphrase this, we get two asymptotes that the branch gets closer to, but not exactly crossing. On the other hand, if $\cos(\theta) > 1/e$, then $1 - e\cos(\theta) < 0$, so r must be negative. This means we get another branch of the curve leftside of the directrix.

So far, the directrix is vertically left of the focus. More generalized version of this occurs when we rotate this by θ_0 counterclockwise. Have a look at the following figure.

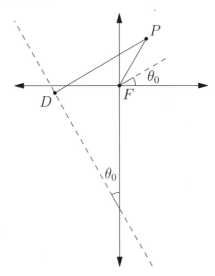

If we simply rotate it counter-clockwise about θ_0, we get the previous case. Hence, the general formula must be $r = \dfrac{ed}{1 - e\cos(\theta - \theta_0)}$ where θ is the angle formed by the x-axis and $\overline{PF}$.

Hence, $PD = r\cos(\theta - \theta_0) + d$ and $PF = r$ to retrieve the general formula. Extending this idea bit more, we get the following forms, assuming that the original expression has the directrix at $x = -d$.

- if $\theta_0 = 90°$, then $r = \dfrac{ed}{1 - e\cos(\theta - 90°)} = \dfrac{ed}{1 - e\sin(\theta)}$ with the directrix at $y = -d$.

- if $\theta_0 = 180°$, then $r = \dfrac{ed}{1 - e\cos(\theta - 180°)} = \dfrac{ed}{1 + e\cos(\theta)}$ with the directrix at $x = d$.

- if $\theta_0 = 270°$, then $r = \dfrac{ed}{1 - e\cos(\theta - 270°)} = \dfrac{ed}{1 + e\sin(\theta)}$ with the directrix at $y = d$.

13.

(a) Change $r = \dfrac{2}{2 - 3\cos(\theta)}$ into rectangular form.

(b) Convert $r = \dfrac{2}{4 + 3\cos(\theta)}$ into rectangular form.

(c) Convert $r = \dfrac{3}{1 + 2\sin(\theta)}$ into rectangular form.

(d) Convert $r = \dfrac{1}{1 - 4\cos(\theta)}$ into rectangular form.

14. If $(x + 1) = \frac{1}{4}y^2$, rotate it $90°$ clockwise to retrieve $r = f(\theta)$. Find $f(\theta)$.

15. Given an ellipse satisfying $\dfrac{(x-1)^2}{4} + \dfrac{y^2}{3} = 1$, find its directrix located leftside of the center.

When we graph polar equation, it is important to notice about symmetries. Let's refresh our memories on line symmetries or point symmetry at Cartesian plane.

- (a, b) is symmetric to $(a, -b)$ with respect to the x-axis.

- (a, b) is symmetric to $(-a, b)$ with respect to the y-axis.

- (a, b) is symmetric to $(-a, -b)$ with respect to the Origin.

Back to our main topic, we say that the graph of $r = f(\theta)$ is the set of points (r, θ) satisfying $r = f(\theta)$. As we figure out shapes of the graph of elementary functions, this section deals with special types of relations that require symmetries. As it is convenient to use line symmetries or point symmetries for Cartesian equations, it must be convenient to use some symmetries with respect to the x-axis, y-axis, and the Origin.

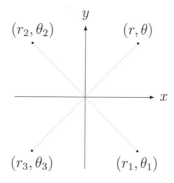

Given (r, θ), then we may find out the following equations

$$(r_1, \theta_1) = (r, -\theta) \text{ or } (-r, \pi - \theta)$$
$$(r_2, \theta_2) = (r, \pi - \theta) \text{ or } (-r, -\theta)$$
$$(r_3, \theta_3) = (r, \theta - \pi) \text{ or } (-r, \theta)$$

Therefore, assuming $r > 0$, we may get the following results.

- If the polar graph of $r = f(\theta)$ is reflected about the x-axis, then

$$r = f(-\theta) \text{ or } r = -f(\pi - \theta)$$

- If the polar graph of $r = f(\theta)$ is reflected about the y-axis, then

$$r = -f(-\theta) \text{ or } r = f(\pi - \theta)$$

- If the polar graph of $r = f(\theta)$ is reflected about the Origin, then

$$r = -f(\theta) \text{ or } r = f(\pi + \theta)$$

On the other hand, if the graph satisfies

1. $f(\theta) = f(-\theta)$ or $f(\theta) = -f(\pi - \theta)$, then the graph of $r = f(\theta)$ is symmetric with respect to the x-axis. For example, $r = \cos(\theta)$ is symmetric about the x-axis.

2. $f(\theta) = -f(-\theta)$ or $f(\theta) = f(\pi - \theta)$, then the graph of $r = f(\theta)$ is symmetric with respect to the y-axis. For example, $r = \sin(\theta)$ is symmetric about the x-axis.

3. $f(\theta) = f(\pi + \theta)$, then the graph of $r = f(\theta)$ is symmetric with respect to the Origin.

Example

Find the polar equation of the graph of $r = 2 - 2\cos\theta$ reflected about the x-axis.

> **Solution**
> We get either $r = 2 - 2\cos(-\theta)$ or $r = -2 + 2\cos(\pi - \theta)$.

16.

(a) Determine whether the graph of $r = 2 - 2\cos(\theta)$ is symmetric with respect to the x-axis.

(b) Determine whether the graph of $r = 3 - 2\sin(\theta)$ is symmetric with respect to the y-axis.

Other than conic sections, we look at four common polar equations that often appear in Calculus. First, we would like to see how **cardioid** looks like. A polar equation of form $r = a \pm a \cos\theta$ or $r = a \pm a \sin\theta$ where $a > 0$. Peculiar trait about cardioid is that

- the graph passes through the pole.

- it looks like a heart.

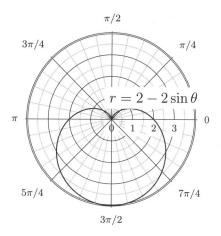

17. Graph $r = 3 + 3\cos\theta$.

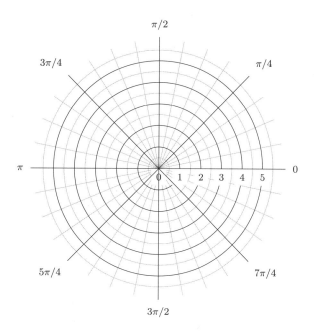

Limacons are divided into two types. The first type has a graph that does not pass through the pole. In other words, r is always greater than 0. The second type has a graph that passes through the pole, creating a small loop. The first type is a limacon <u>without inner loop</u>. The equation is given by

$$r = a \pm b \cos \theta \text{ or } r = a \pm b \sin \theta$$

where $0 < |b| < |a|$. On the other hand, the second type is a limacon <u>with an inner loop</u>, whose equation is given by

$$r = a \pm b \cos \theta \text{ or } r = a \pm b \sin \theta$$

where $0 < |a| < |b|$. Have a look at the two figures.

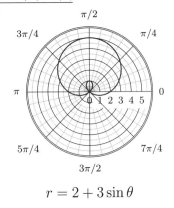

$$r = 2 + 3 \sin \theta$$

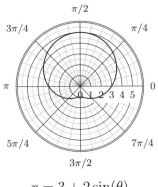

$$r = 3 + 2 \sin(\theta)$$

Graphing limacons having $\cos(\theta)$, we simply trace the graph with θ from 0 to π, and reflect that image about the x-axis. On the other hand, graphing limacons having $\sin(\theta)$, we trace it with θ from $-\dfrac{\pi}{2}$ to $\dfrac{\pi}{2}$, and reflect that image about the y-axis.

18. Graph $r = 4 - 2 \cos \theta$.

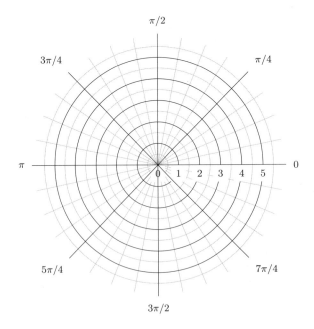

The third type is a **rose**. A polar equation $r = a\cos(n\theta)$ or $r = a\sin(n\theta)$ where $a > 0$ is a rose. If n is even, there are $2n$ petals. If n is odd, there are n petals. When we sketch a rose, we care about when

- r reaches either its maximum or minimum, telling us where the tip of a petal is located.

- r reaches 0, telling us the size of a petal.

Let's look at the graph of $r = 5\cos(2\theta)$.

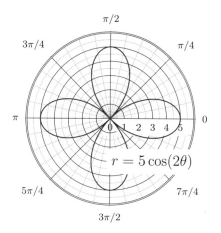

19. Graph $r = 3\sin(3\theta)$.

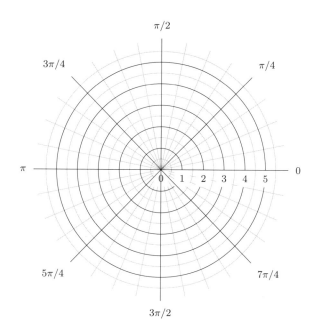

Last type is a **lemniscate**, introduced by Bernoulli, currently used as a symbol for infinity. A polar equation $r^2 = a^2 \sin(2\theta)$ or $r^2 = a^2 \cos(2\theta)$ where $a > 0$ is a lemniscate. Remember that $r^2 \geq 0$ so that θ must be restricted. The graph looks like a propeller. When we plot a lemniscate, remember two things.

1) $r^2 = a^2 \sin(2\theta)$ where $0 \leq \theta \leq \dfrac{\pi}{2}$

2) $r^2 = a^2 \cos(2\theta)$ where $-\dfrac{\pi}{4} \leq \theta \leq \dfrac{\pi}{4}$

Because $r^2 \geq 0$, $a^2 \sin(2\theta)$ must be non-negative. Likewise, $a^2 \cos(2\theta)$ must be non-negative. This is the reason why there is a domain restriction.

Let's look at the graph of $r^2 = 4 \sin(2\theta)$.

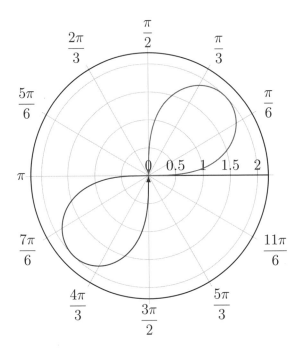

When we graph a lemniscate, we use the fact that r^2 makes two possible directions. In other words, when we graph portions of lemniscate in the first quadrant, then we draw the other corresponding graph in the third quadrant, due to the origin symmetry. As one can check from the figure above, southwest portions of the graph are automatically copied to produce northeast portions of the graph.

⬛ Core Skill Practice 14

01.
Find polar coordinates of a point whose rectangular coordinates are $(-1, \sqrt{3})$.

02.
Convert the following rectangular coordinates to polar coordinates.

(a) $(0, -3)$ (b) $(5, 5)$ (c) $(-2\sqrt{3}, -6)$

03.
Convert the following polar coordinates to rectangular coordinates.

(a) $(8, \frac{\pi}{3})$ (b) $(12, -120°)$ (c) $(4\sqrt{2}, \frac{7\pi}{4})$

04.

Transform $r = 3\cos\theta$ into rectangular form.

05.

Find a polar form of $3x = \dfrac{2}{y}$.

06.

Change $r = 2\sin\theta$ into rectangular form.

07.

Transform $r = \dfrac{4}{4\cos\theta - 6\sin\theta}$ into rectangular form.

08.

Change the following polar equations into rectangular equations.

(a) $r\cos(\theta) = 2$

(b) $r = \dfrac{4}{2\cos\theta - \sin\theta}$

(c) $r = \cos\theta$

09.

Change the following polar equations into rectangular equations.

(a) $r\sin\theta = 3$

(b) $r(1 - \cos\theta) = 2$

(c) $r^2\cos(2\theta) = 4$

10.

Change $x^2 + y^2 - 2y = 0$ into polar equation.

11.

Change $r \sin \theta = 3$ into rectangular equation.

12.

Find the eccentricity of $r = \dfrac{2}{2 + 3\cos(\theta)}$.

13. Graph the equation $r = 3 - \sin(\theta)$.

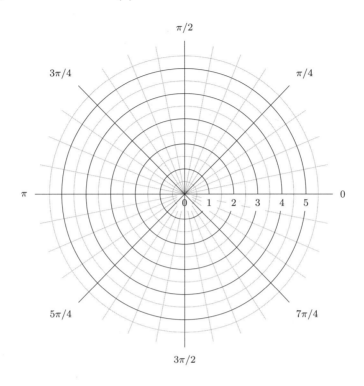

14. Graph the equation $r = 2 + 2\cos(\theta)$.

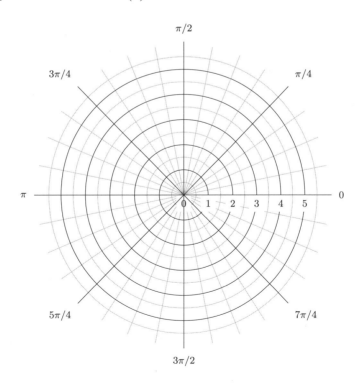

15. Graph the equation $r = 2\sin(3\theta)$.

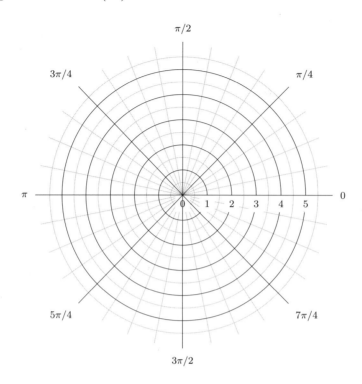

💡 Solution to Core Skill Practice 14

01. $(r, \theta) = (2, \dfrac{2\pi}{3})$.

02.
(a) $(3, \dfrac{3\pi}{2})$ (b) $(5\sqrt{2}, \dfrac{\pi}{4})$ (c) $(4\sqrt{3}, \dfrac{4\pi}{3})$

03.
(a) $(4, 4\sqrt{3})$ (b) $(-6, -6\sqrt{3})$ (c) $(4, -4)$

04. $(x - \dfrac{3}{2})^2 + y^2 = \dfrac{9}{4}$

05. $r^2 = \dfrac{2}{3\cos\theta\sin\theta}$

06. $x^2 + (y - 1)^2 = 1$

07. $y = \dfrac{2}{3}x - \dfrac{2}{3}$

08.
(a) $x = 2$ (b) $y = 2x - 4$ (c) $(x - \dfrac{1}{2})^2 + y^2 = \dfrac{1}{4}$

09.
(a) $y = 3$ (b) $y^2 = 4(x + 1)$ (c) $x^2 - y^2 = 4$

10. $r = 2\sin(\theta)$

11. $y = 3$

12. $e = \dfrac{3}{2}$

13.

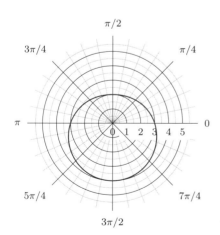

14.

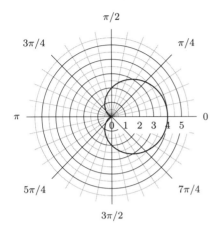

15.

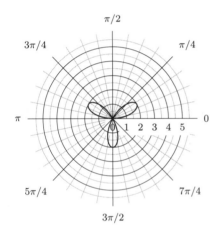

TOPIC
9

Matrices

A matrix is an array of numbers in a row and column.

$$\begin{bmatrix} x_{11} & x_{12} & x_{13} \\ x_{21} & x_{22} & x_{23} \end{bmatrix}$$

This is a matrix of 2×3 dimension where x_{ij} is the entry at ith row and jth column. Given a set of matrices of $m \times n$ dimension, we have the following properties satisfied. Given two matrices of equal dimension, we add or subtract elements at each entry. Note that only two matrices of equal dimension can be added or subtracted.

$$\begin{bmatrix} x_{11} & x_{12} & x_{13} \\ x_{21} & x_{22} & x_{23} \end{bmatrix} \pm \begin{bmatrix} y_{11} & y_{12} & y_{13} \\ y_{21} & y_{22} & y_{23} \end{bmatrix} = \begin{bmatrix} x_{11} \pm y_{11} & x_{12} \pm y_{12} & x_{13} \pm y_{13} \\ x_{21} \pm y_{21} & x_{22} \pm y_{22} & x_{23} \pm y_{23} \end{bmatrix}$$

Additionally, there is a scalar multiple, which is the product of a scalar (=a numeral magnitude) and a matrix of any dimension.

$$k \begin{bmatrix} y_{11} & y_{12} & y_{13} \\ y_{21} & y_{22} & y_{23} \end{bmatrix} = \begin{bmatrix} ky_{11} & ky_{12} & ky_{13} \\ ky_{21} & ky_{22} & ky_{23} \end{bmatrix}$$

There is also a matrix that is called a null matrix or zero matrix whose all entries are 0s, which is denoted by $O_{m \times n}$. For instance, a 2×2 zero matrix is given by

$$O_{2 \times 2} = \begin{bmatrix} 0 & 0 \\ 0 & 0 \end{bmatrix}$$

which satisfies $A_{2 \times 2} \pm O_{2 \times 2} = A_{2 \times 2}$ for any matrix $A_{2 \times 2}$.

Example

Evaluate $3 \begin{bmatrix} 3 & 2 & 4 \\ 1 & 2 & -1 \end{bmatrix} + 2 \begin{bmatrix} 2 & 0 & 1 \\ -1 & 3 & 2 \end{bmatrix}$

Solution

$$\begin{bmatrix} 9 & 6 & 12 \\ 3 & 6 & -2 \end{bmatrix} + \begin{bmatrix} 4 & 0 & 2 \\ -2 & 6 & 4 \end{bmatrix} = \begin{bmatrix} 13 & 6 & 14 \\ 1 & 12 & 2 \end{bmatrix}$$

The tranpose of a matrix A is written as A^T and is found by interchanging rows and columns. Since we are interchanging rows and columns, dimension also gets switched.

Example

If $A = \begin{bmatrix} 3 & 1 \\ 2 & 6 \end{bmatrix}$, then $A^T = \begin{bmatrix} 3 & 2 \\ 1 & 6 \end{bmatrix}$.

Condition for Matrix Multiplication

Given two matrices $A_{m \times n}$ and $B_{n \times p}$, then AB has the dimension of $m \times p$. Otherwise, it is impossible to perform multiplication. Also, matrix multiplication is not commutative.

$$A_{m \times n} B_{n \times p} = AB_{m \times p} \neq BA \text{(which does not exist in this case)}$$

Row-Column Multiplication : Dot Product

The usual rule for row-column multiplication can be illustrated by the following. Suppose there is a $X_{2 \times 3}, Y_{3 \times 2}$ and $XY_{2 \times 2}$. Then, the entry at the ith row and the jth column of XY is computed by a dot product of the ith row of X and the jth column of Y. For instance, the dot product of the first row of X and the first column of Y results in z_{11}.

Row-Column Multiplication

$$\begin{bmatrix} x_{11} & x_{12} & x_{13} \\ \cdots & \cdots & \cdots \end{bmatrix} \times \begin{bmatrix} y_{11} & \cdots \\ y_{21} & \cdots \\ y_{31} & \cdots \end{bmatrix} = \begin{bmatrix} z_{11} & \cdots \\ \cdots & \cdots \end{bmatrix}$$

where $z_{11} = \begin{bmatrix} x_{11} & x_{12} & x_{13} \end{bmatrix} \cdot \begin{bmatrix} y_{11} & y_{21} & y_{31} \end{bmatrix} = x_{11}y_{11} + x_{12}y_{21} + x_{13}y_{31} = \sum_{k=1}^{n} x_{1k}y_{k1}$

1. Find the product $\begin{bmatrix} 2 & 3 \\ 3 & -2 \end{bmatrix} \begin{bmatrix} 3 & -1 \\ 2 & 4 \end{bmatrix}$ using the row-column multiplication.

Column-Row Multiplication : Linear Combination

This is more useful than row-column multiplication when the size of the matrices becomes large. Look at the following illustration.

$$\begin{bmatrix} 3 & 1 \\ 2 & 4 \end{bmatrix} \begin{bmatrix} 2 \\ 3 \end{bmatrix} = \begin{bmatrix} c_{11} \\ c_{21} \end{bmatrix}$$

Instead of looking at row-column multiple, we are looking at a linear combination of the column vectors of the first matrix. To be specific,

$$\begin{bmatrix} c_{11} \\ c_{21} \end{bmatrix} = 2 \begin{bmatrix} 3 \\ 2 \end{bmatrix} + 3 \begin{bmatrix} 1 \\ 4 \end{bmatrix}$$

$$= \begin{bmatrix} 9 \\ 16 \end{bmatrix}$$

2. Compute $\begin{bmatrix} 3 & 4 & 5 \\ 1 & 2 & 3 \\ 0 & 1 & 2 \end{bmatrix} \begin{bmatrix} 1 & 2 & 3 \\ 2 & 4 & 1 \\ 0 & 4 & 5 \end{bmatrix}$ using the column-row multiplication.

First of all, inverse matrices exist only for square matrices of $n \times n$ dimension. In Precalculus, mainly two types of matrices are covered. For a 2×2 matrix $A = \begin{bmatrix} a & b \\ c & d \end{bmatrix}$, the inverse of A is given by

$$A^{-1} = \frac{1}{ad - bc} \begin{bmatrix} d & -b \\ -c & a \end{bmatrix}$$

where $ad - bc$ is called the determinant, also known as $\det(A)$. The determinant of a 2×2 matrix is associated with the area of triangle formed by the two different vectors, given by the two column vectors. If the determinant is 0, then A is not invertible and called singular. If not, then A is invertible.

- $ad - bc = 0 \iff A^{-1}$ does not exist.

- $ad - bc \neq 0 \iff A^{-1}$ exists.

For a 3×3 matrix, we should learn how to compute minors, cofactor and transpose in order to find the inverse matrix. In fact, this process will work for all square matrices, even for 2×2 matrices. The following subsections will illustrate how to compute them one by one.

The determinant of a 3×3 matrix $A = \begin{bmatrix} a & b & c \\ d & e & f \\ g & h & i \end{bmatrix}$ is equal to

$$\det A = a \begin{vmatrix} e & f \\ h & i \end{vmatrix} - b \begin{vmatrix} d & f \\ g & i \end{vmatrix} + c \begin{vmatrix} d & e \\ g & h \end{vmatrix}$$

For instance, the determinant of $\begin{bmatrix} 3 & 0 & 2 \\ 2 & 0 & -2 \\ 0 & 1 & 1 \end{bmatrix}$ is equal to

$$3 \begin{vmatrix} 0 & -2 \\ 1 & 1 \end{vmatrix} - 0 \begin{vmatrix} 2 & -2 \\ 0 & 1 \end{vmatrix} + 2 \begin{vmatrix} 2 & 0 \\ 0 & 1 \end{vmatrix} = 10$$

Minor

Minor is a submatrix A_{ij} of determinants formed by crossing out the ith row and jth column.

Cofactor

Matrix of cofactor assigns $+, -$ signs for each entry. For 3×3 matrices, the cofactor matrix is given by

$$\begin{bmatrix} + & - & + \\ - & + & - \\ + & - & + \end{bmatrix}$$

For + entry, we do not change the sign of the entry. For − entry, on the other hand, we should change the sign of the entry by putting − sign in front of each value.

Adjugate

As we covered it earlier, there is a matrix A^T associated to every matrix A, which is known as transposition. The matrix of adjugate is to transpose a given matrix A, which is written as A^T, and it can be found by interchanging the rows and columns, i.e., $A_{ij} \to A_{ji}$.

$$A = \begin{bmatrix} a & b \\ c & d \end{bmatrix}$$

$$A^T = \begin{bmatrix} a & c \\ b & d \end{bmatrix}$$

3.

(a) Find the matrix of minors of $\begin{bmatrix} 3 & 0 & 2 \\ 2 & 0 & -2 \\ 0 & 1 & 1 \end{bmatrix}$.

(b) Find the matrix of cofactor of $\begin{bmatrix} 2 & 2 & 2 \\ -2 & 3 & 3 \\ 0 & -10 & 0 \end{bmatrix}$.

(c) Find the transpose of $\begin{bmatrix} 2 & -2 & 2 \\ 2 & 3 & -3 \\ 0 & 10 & 0 \end{bmatrix}$.

As one could guess, the inverse matrix of $\begin{bmatrix} 3 & 0 & 2 \\ 2 & 0 & -2 \\ 0 & 1 & 1 \end{bmatrix}$ is equal to

$$\frac{1}{10} \begin{bmatrix} 2 & 2 & 0 \\ -2 & 3 & 10 \\ 2 & -3 & 0 \end{bmatrix}$$

The identity matrix I is given by $I = \begin{bmatrix} 1 & 0 \\ 0 & 1 \end{bmatrix}$ or $I = \begin{bmatrix} 1 & 0 & 0 \\ 0 & 1 & 0 \\ 0 & 0 & 1 \end{bmatrix}$ such that for $A_{2 \times 2}$ or $A_{3 \times 3}$,

$$AI = IA = A$$

Furthermore, the product of A and A^{-1} results in the identity matrix.

$$AA^{-1} = A^{-1}A = I$$

Example

Compute $\dfrac{1}{10} \begin{bmatrix} 2 & 2 & 0 \\ -2 & 3 & 10 \\ 2 & -3 & 0 \end{bmatrix} \begin{bmatrix} 3 & 0 & 2 \\ 2 & 0 & -2 \\ 0 & 1 & 1 \end{bmatrix}$

Solution
As you have guessed, the product of two inverse matrices result in $I_{3 \times 3}$, the identity matrix.

4. Find the determinant of $\begin{bmatrix} 4 & 1 & 2 \\ 0 & 1 & 3 \\ 2 & 2 & 1 \end{bmatrix}$.

5. For $A = \begin{bmatrix} -2 & p-1 & 1 \\ 2 & 1 & p+1 \\ -4 & 2 & p-2 \end{bmatrix}$, solve the following questions.

(a) Find an expression for $\det A$ in terms of p.

(b) Given that $\det A = 8$, find the possible values of p.

6. Given that $C = \begin{bmatrix} 1 & 2 & 3 \\ 2 & 1 & 0 \\ 1 & -3 & -2 \end{bmatrix}$, find the matrix C^{-1}.

9.3 Elementary Matrices

Since we saw how row-column multiplication works, we would like to see whether there are similar processes using rows. In fact, Gauss came up with elementary row operations, which we have used in solving the system of equations taught in Algebra 1 and Algebra 2. What are elementary row operations? What is an *elementary* matrix? There are three elementary matrices that switch rows, multiply a row by a constant, and add a scalar multiple of a row to another row. All these operations do not change the solution to the system of equations. Though this seems overshot for Precalculus, this is required to completely understand the row-reduced echelon form that some highschool students learn in Honors Precalculus course. Lastly, we will have a solid understanding of what Gauss saw when he solved the system of linear equations.

- Switching rows

- Multiplying a scalar

- Adding/subtracting rows

7.

(a) Compute $\begin{bmatrix} 0 & 1 & 0 \\ 1 & 0 & 0 \\ 0 & 0 & 1 \end{bmatrix} \begin{bmatrix} a & b & c \\ d & e & f \\ h & i & j \end{bmatrix}$.

(b) Compute $\begin{bmatrix} 1 & 0 & 0 \\ 0 & 2 & 0 \\ 0 & 0 & 1 \end{bmatrix} \begin{bmatrix} a & b & c \\ d & e & f \\ h & i & j \end{bmatrix}$.

(c) Compute $\begin{bmatrix} 1 & 0 & 0 \\ -1 & 1 & 0 \\ 0 & 0 & 1 \end{bmatrix} \begin{bmatrix} a & b & c \\ d & e & f \\ h & i & j \end{bmatrix}$.

Multiplying a set of elementary matrices in front of a given matrix does not change the *nature* of the expression. In other words, it does not put a new piece of information into our computation. This is extremely convenient, since we can safely solve a matrix equation by multiplying tons of elementary matrices in front to simplify the given expression. This is called Gauss-Jordan elimination method, and we shall investigate it soon. That being written, we would like to the order of multiplying elementary matrices also matters.

8. Compute

(a)

$$\begin{bmatrix} 2 & 0 & 0 \\ 0 & 0 & 2 \\ 1 & 0 & 0 \end{bmatrix} \begin{bmatrix} 1 & 1 & 0 \\ 0 & -1 & 1 \\ 1 & 0 & 1 \end{bmatrix} \begin{bmatrix} 2 & 1 & 1 \\ 1 & -3 & -1 \\ -2 & 3 & 1 \end{bmatrix}$$

(b)

$$\begin{bmatrix} 1 & 1 & 0 \\ 0 & -1 & 1 \\ 1 & 0 & 1 \end{bmatrix} \begin{bmatrix} 2 & 0 & 0 \\ 0 & 0 & 2 \\ 1 & 0 & 0 \end{bmatrix} \begin{bmatrix} 2 & 1 & 1 \\ 1 & -3 & -1 \\ -2 & 3 & 1 \end{bmatrix}$$

Gauss-Jordan elimination method gives us a systematic way of solving systems of linear equations. This is the most useful tool we have for Linear Algebra. First, we have to learn something called augmented matrix of a system of equations. The augmented matrix has its entries as the coefficients of all the variables and the constants of the right-hand side.

For instance, the system of equations $\begin{cases} x + 2y + z = 3 \\ x - y + 3z = 4 \end{cases}$ turns into

$$\begin{bmatrix} 1 & 2 & 1 & | & 3 \\ 1 & -1 & 3 & | & 4 \end{bmatrix}$$

Now, Gauss-Jordan elimination takes a matrix and puts it into a specialized form known as *reduced echelon form*, using elementary row operations. The key idea here is to eliminate entries above and below *pivots*.

Look at the following augmented matrix.

$$\begin{bmatrix} 2 & 1 & 1 & | & 4 \\ -2 & 2 & 3 & | & 3 \\ 1 & 2 & 4 & | & 7 \end{bmatrix}$$

The pivot position(the first nonzero entry) of the first row is "2" in the first column. The first thing we do is to turn every entry below this pivot position into 0. We do this by the following row operations.

$$I + II \to II$$
$$-I + 2III \to III$$

where roman numerals indicate the row numbers and an expression like $-I + 2III \to III$ means multiplying the first row by -1 and the third row by 2 and putting the result into the third row.

9. Compute the following row operations.

$$\begin{bmatrix} 1 & 0 & 0 \\ 1 & 1 & 0 \\ -1 & 0 & 2 \end{bmatrix} \begin{bmatrix} 2 & 1 & 1 & | & 4 \\ -2 & 2 & 3 & | & 3 \\ 1 & 2 & 4 & | & 7 \end{bmatrix}$$

The two row operations gave us the matrix

$$\begin{bmatrix} 2 & 1 & 1 & | & 4 \\ 0 & 3 & 4 & | & 7 \\ 0 & 3 & 7 & | & 10 \end{bmatrix}$$

Now, the middle 3 in the second row turns a new pivot position. We will eliminate entries above and below the pivot position by the following row operations.

$$-3I + II \rightarrow I$$
$$-II + III \rightarrow III$$

10. Compute the following row operations.

$$\begin{bmatrix} -3 & 1 & 0 \\ 0 & 1 & 0 \\ 0 & -1 & 1 \end{bmatrix} \begin{bmatrix} 2 & 1 & 1 & | & 4 \\ 0 & 3 & 4 & | & 7 \\ 0 & 3 & 7 & | & 10 \end{bmatrix}$$

The previous row operation results in

$$\begin{bmatrix} -6 & 0 & 1 & | & -5 \\ 0 & 3 & 4 & | & 7 \\ 0 & 0 & 3 & | & 3 \end{bmatrix}$$

This time, we change our pivot position to 3 in the last row. We will perform another set of row operations to get rid of 1 and 4 in the third column.

$$\frac{1}{3}III \rightarrow III$$

$$-3II + 4III \rightarrow II$$

$$I - \frac{1}{3}III \rightarrow I$$

11. Compute the following row operations.

$$\begin{bmatrix} 1 & 0 & -\dfrac{1}{3} \\ 0 & -3 & 4 \\ 0 & 0 & \dfrac{1}{3} \end{bmatrix} \begin{bmatrix} -6 & 0 & 1 & | & -5 \\ 0 & 3 & 4 & | & 7 \\ 0 & 0 & 3 & | & 3 \end{bmatrix}$$

The resulting augmented matrix is given by

$$\left[\begin{array}{ccc|c} -6 & 0 & 0 & -6 \\ 0 & -9 & 0 & -9 \\ 0 & 0 & 1 & 1 \end{array}\right]$$

The last row operation will change the coefficient of the augmented matrix into 1's.

$$-\frac{1}{6}I \to I$$
$$-\frac{1}{9}II \to II$$

The elementary matrix associated to this row operation is

$$\left[\begin{array}{ccc} -\dfrac{1}{6} & 0 & 0 \\ 0 & -\dfrac{1}{9} & 0 \\ 0 & 0 & 1 \end{array}\right]$$

If we multiply this matrix to the given augmented matrix, we get the final row-reduced echelon form,

$$\left[\begin{array}{ccc|c} 1 & 0 & 0 & 1 \\ 0 & 1 & 0 & 1 \\ 0 & 0 & 1 & 1 \end{array}\right]$$

The given matrix tells us that the system of equation has its corresponding coefficients such that

$$\begin{cases} 1x + 0y + 0z & = 1 \\ 0x + 1y + 0z & = 1 \\ 0x + 0y + 1z & = 1 \end{cases}$$

Hence, $x = 1, y = 1, z = 1$.

As a parting shot for Precalculus, this section will provoke readers to continue studying this subject called Linear Algebra. For simplicity, we will restrict our attention to 2×2 matrices. **Eigenvalue** is a real value λ such that $A\mathbf{x} = \lambda\mathbf{x}$. It may sound difficult when matrix multiplication equals a scalar multiple for some $\mathbf{x}$, but this is extremely useful in Linear Algebra because such transformation of axes allows us to change dirty expressions into neat forms. Here is a better explanation with figures. When we multiply a matrix A to $[x_1, x_2]$, we simply send a point $[x_1, x_2]$ to another one by multiplying a 2×2 matrix A to $[x_1, x_2]$.

When we say we are looking for eigenvalue, we are looking for some $[x_1, x_2]$ such that $A \cdot \mathbf{x}$ is a scalar multiple of $[x_1, x_2]$, i.e., $\lambda\mathbf{x}$, and such λ is a real number. In the following figures, the left figure shows the result by matrix multiplication that sends $\mathbf{x}$ to $A\mathbf{x}$ or $\mathbf{y}$ to $A\mathbf{y}$. On the other hand, the right figure shows the scalar multiple of $\mathbf{x}$. As shown in the right figure, there could be two different values of λ.

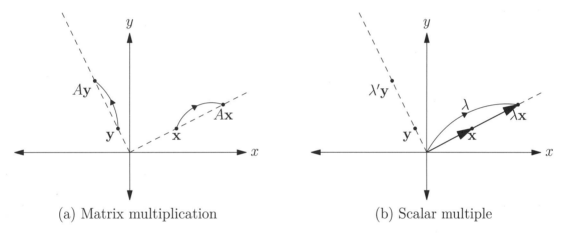

(a) Matrix multiplication　　　　　　(b) Scalar multiple

12. If $A = \begin{bmatrix} 1 & 2 \\ 3 & 4 \end{bmatrix}$, find two values of λs such that there exists a non-trivial $\mathbf{x} = \begin{bmatrix} x_1 \\ x_2 \end{bmatrix}$ such that $A\mathbf{x} = \lambda\mathbf{x}$.

01.
Compute

$$\begin{bmatrix} 1 & 1 & 0 \\ 0 & -1 & 1 \\ 1 & 0 & 1 \end{bmatrix} \begin{bmatrix} 2 & 1 & 1 \\ 1 & -3 & -1 \\ -2 & 3 & 1 \end{bmatrix}$$

02.
Compute

$$\begin{bmatrix} 0 & 0 & 1 \\ 1 & 0 & 0 \\ 0 & 1 & 0 \end{bmatrix} \begin{bmatrix} 2 & 1 & 1 \\ 1 & -3 & -1 \\ -2 & 3 & 1 \end{bmatrix}$$

03.

Compute

$$\begin{bmatrix} 2 & 0 & 0 \\ 0 & 0 & -3 \\ 0 & 2 & 0 \end{bmatrix} \begin{bmatrix} 2 & 1 & 1 \\ 1 & -3 & -1 \\ -2 & 3 & 1 \end{bmatrix}$$

04.

Compute

$$\begin{bmatrix} 2 & -3 & 0 \\ 3 & 1 & 0 \\ 0 & -2 & 1 \end{bmatrix} \begin{bmatrix} 2 & 1 & 1 \\ 1 & -3 & -1 \\ -2 & 3 & 1 \end{bmatrix}$$

05.

Find the sum of all values of λ such that

$$\begin{bmatrix} 1 & 3 \\ 2 & 1 \end{bmatrix} \begin{bmatrix} x \\ y \end{bmatrix} = \lambda \begin{bmatrix} x \\ y \end{bmatrix}$$

has other solution that $x = 0$ and $y = 0$.

06.

Use Gauss-Jordan elimination method to solve

$$\begin{cases} 3x + 4y = 11 \\ 8x - 6y = -4 \end{cases}$$

07.

Use Gauss-Jordan elimination method to solve

$$\begin{cases} 3x - 2y + z = 2 \\ x + 3y + 2z = 6 \\ 6x + y - z = 6 \end{cases}$$

08.

If $f(x) = ax^2 + bx + c$ where $f(1) = 8$, $f(-2) = 2$, and $f(-3) = 4$.

09.

Find the inverse matrix of

$$\begin{bmatrix} 1 & 3 & 1 \\ 2 & 0 & 3 \\ -1 & 2 & 1 \end{bmatrix}$$

10.

Find the determinant of

$$\begin{bmatrix} 0 & 2 & 1 \\ 1 & 4 & 1 \\ 3 & -1 & 0 \end{bmatrix}$$

☀ Solution to Core Skill Practice 15

01. $\begin{bmatrix} 3 & -2 & 0 \\ -3 & 6 & 2 \\ 0 & 4 & 2 \end{bmatrix}$

02. $\begin{bmatrix} -2 & 3 & 1 \\ 2 & 1 & 1 \\ 1 & -3 & -1 \end{bmatrix}$

03. $\begin{bmatrix} 4 & 2 & 2 \\ 6 & -9 & -3 \\ 2 & -6 & -2 \end{bmatrix}$

04. $\begin{bmatrix} 1 & 11 & 5 \\ 7 & 0 & 2 \\ -4 & 9 & 3 \end{bmatrix}$

05. 2

06. $(x, y) = (1, 2)$

07. $(x, y, z) = (1, 1, 1)$

08. $(a, b, c) = (1, 3, 4)$

09. $\begin{bmatrix} 6/17 & 1/17 & -9/17 \\ 5/17 & -2/17 & 1/17 \\ -4/17 & 5/17 & 6/17 \end{bmatrix}$

10. -7

◆ 교재를 마치며

이 교재를 작성하며 제자들의 피드백을 받을 때, 가장 듣기 좋은 말은 "내가 학창 시절에 이런 교재가 있었으면 좋았을 것을….."이라는 말이었습니다. 미국 수학 교과 과정 중 가장 중요하다고 볼 수 있는 Precalculus 교재를 집필하면서 염두에 두었던 부분은 상위권으로 도약하려고 하는 학생들에게 반드시 도움이 되었으면 하는 점과 이 교재로 수업하는 선생님도 교재의 내용을 보며 수학을 재미있게 가르칠 수 있었으면 하는 점이었습니다.

Function 부분을 설명할 때는 그래프를 그려내는 능력을 길러주고 싶었고, 특히 합성함수를 바라보는 관점을 길러주길 희망하며 교재를 썼습니다. Algebra 2에서 배운 기본 내용을 조금 더 응용해서, 함수에 대한 개념들을 재미있게 응용할 수 있으면 학생들에게 정말 도움이 되겠다 싶은 마음으로 교재를 썼습니다.

또한, Trigonometry 부분을 설명할 때는 그래프 뿐만 아니라, 정수론과의 응용, 가우스 함수와의 합성을 포함하며, 교과 과정에서 쉬이 다루지 않는 부분을 섞어보면서, 학생들에게 조금 더 깊은 생각을 하길 바라는 마음으로 집필해보았습니다. 뿐만 아니라, 교과 개념이 유기적으로 연결될 수 있도록 역사적인 관점도 넣어보았습니다. 특히, Descartes, Viete, Newton, De Moivre 등 우리가 세계사를 공부할 때, 한번쯤 들어 볼법한 역사적 인물들에 대한 내용을 간간히 포함하다 보니, 이과 학생 혹은 문과 학생의 접근을 용이하게 할 수 있는 Precalculus 교재를 집필할 수 있었습니다.

이 교재를 통해, 학생들이 학교에서 좋은 성적을 거둘 뿐 아니라, 미국 고교 수학 과정의 핵심을 바라보는 안목을 기르길 바라는 마음으로 끝맺음합니다. 학생 여러분, 즐겁게 공부하느라 수고했고 고맙습니다.

Solution
Manual

Solutions for Topic **1**

1 Let point F lies on the intersection between the y-axis and the segment $\overline{AB}$. Then, the line equation that contains $\overline{AB}$ is $y = -\dfrac{1}{3}x + \dfrac{7}{3}$. Hence, $F = (0, \dfrac{7}{3})$. Therefore, $S(\triangle ABD) = S(\triangle BFD) + S(\triangle AFD) = \dfrac{13}{3} + \dfrac{13}{6} = \dfrac{13}{2}$. Thus, the area is $\dfrac{13}{2}$.

2 According to internal section formula, we get $B = \left(\dfrac{2(6) + 3(1)}{5}, \dfrac{4(3) + 2(-1)}{5} \right) = (3, 2)$. Instead of using internal segment / external segment formula on 2-dimension, consider two coordinates separately.

3

(a)

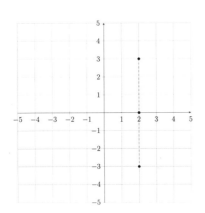

(b)

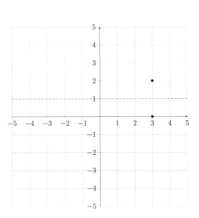

4 If a function is polynomial, we get $f(x) = a_1 x^{2b_1} + a_2 x^{2b_2} + \cdots$. If a function includes absolute value, it must be even. Hence, the correct answer is (C) because $f(-x) = |-x| + 2 = |x| + 2 = f(x)$.

5

(a)

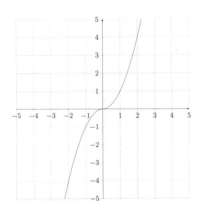

(b)

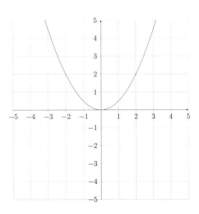

6 (a, b) is the midpoint of $(2, 4)$ and $(-4, 0)$, so we use $a = \dfrac{2 + (-4)}{2} = -1$ and $b = \dfrac{4 + 0}{2} = 2$. Hence, $(a, b) = (-1, 2)$.

7 Since the first line of the question implies that $f(x) = f(-x)$, we know that $f(x)$ is even function. The correct answer is (B) because $(0, 3)$ must be the midpoint between the original point and the reflected point.

8

(a) Since $\dfrac{(-x)^2}{3(-x)^2 + 5| - x|} = \dfrac{x^2}{3x^2 + 5|x|}$, we conclude that the graph is symmetric about the y-axis because $f(-x) = f(x)$.

(b) For every real x, we know that $y > 0$. Hence, the graph is not symmetric about the x-axis because $-f(x) \neq f(x)$.

(c) Since the graph is only defined in the first and second quadrants, the graph is not symmetric about the origin because $-f(-x) \neq f(x)$.

9 Since the x-intercepts are $(\pm 3, 0)$ and y-intercepts are $(0, \pm 2)$, the graph is centered at the origin and symmetric about the line $y = 0$ and $x = 0$. Hence, the answer is (C).

10 Since the point is *interior* of the circle, we get $x^2 + y^2 < 25$. Therefore, the only answer that matches with this is (D).

11 $x^2 + (y - 5)^2 = 144 + 25 = 169$. Then, radius is 13. Thus, $S = \pi r^2 = 169\pi$. The answer is (B).

12 We are looking for the distance between the origin and the line $4x + 3y - 10 = 0$. Hence,

$$
\begin{aligned}
d &= \frac{|4 \cdot 0 + 3 \cdot 0 - 10|}{\sqrt{4^2 + 3^2}} \\
&= \frac{10}{5} \\
&= 2
\end{aligned}
$$

13 The center is located at the perpendicular bisector of the segment connecting $(2, 1)$ and $(6, 5)$. Therefore, the graph has the gradient -1 and passes through $(4, 3)$. Therefore, $x + y = 7$. The only answer choice that satisfies the given condition is (D).

14 Since $y = \dfrac{x^2}{2}$ equals $2y = x^2$, we get $4p = 2$. Hence, the length of latus rectum is 2.

15 Since $y = 2x^2 + 4x - 3$ equals $y = 2(x + 1)^2 - 5$, we get $\dfrac{1}{2}(y + 5) = (x + 1)^2$, we get $p = \dfrac{1}{8}$. Since the center is located at $(-1, -5)$, we need to push it down by $\dfrac{1}{8}$. Hence, the answer is (D).

16 Since $x^2 - 2x + 2y^2 + 4y = 5$, we get $(x-1)^2 + 2(y+1)^2 = 8$, we get

$$(x-1)^2 + 2(y+1)^2 = 8$$
$$\frac{(x-1)^2}{8} + \frac{(y+1)^2}{4} = 1$$

The length of major axis is $2\sqrt{8} = 4\sqrt{2}$.

17 First, switch the original form into $(x-1)^2 - 2(y-1)^2 = 1$ by completing the square. Then, we get $(x-1)^2 - \frac{(y-1)^2}{1/2} = 1$. Hence,

$$y = \pm \frac{\sqrt{1/2}}{1}(x-1) + 1$$
$$= \pm \frac{\sqrt{2}}{2}(x-1) + 1$$

Solutions for Topic **2**

1

(a) $f(2) = 3(2)^2 - 2 = 3(4) - 2 = 12 - 2 = 10.$

(b) $f(\alpha) = 3(\alpha)^2 - 2 = 47$, so $3\alpha^2 = 49$. Hence, $\alpha = \pm\dfrac{7}{\sqrt{3}} = \pm\dfrac{7\sqrt{3}}{3}.$

(c) $f(2x - 1) = 3(2x - 1)^2 - 2 = 3(4x^2 - 4x + 1) - 2 = 12x^2 - 12x + 3 - 2 = 12x^2 - 12x + 1.$

(d) $f(1 - 2x) = f(-(2x - 1)) = f(2x - 1) = 12x^2 - 12x + 1$ because $f(x)$ is even.

2 Let $3x = t$. Then $x = \dfrac{t}{3}$, so $f(t) = 4(\dfrac{t}{3})^2 = 4\dfrac{t^2}{9}$. Therefore, $f(g(x)) = \dfrac{4(g(x))^2}{9}$. Hence, the answer is (C).

3 Since $f(g(x)) = 3(2x - 3) + 2 = 6x - 7$, we have $f(g(2)) = 5$. Therefore, the answer is (C) because $f(g(2)) = 5$.

4 $f(x) \cdot f(-x) = \dfrac{x + 4}{x - 4} \times \dfrac{-x + 4}{-x - 4} = \dfrac{x + 4}{x - 4} \times \dfrac{-(x - 4)}{-(x + 4)} = \dfrac{-1}{-1} = 1$ except $|x| \neq 4$. This is different from a constant function at two points $x = 4$ and $x = -4$. If two functions are different from at least one value of x, we call them distinct.

5 The correct answer is (C) because
$(f + g)(x) = f(x) + g(x) = 3x^2 + 3x - 6 = 3(x^2 + x - 2) = 3(x + 2)(x - 1).$

6

(a) $f(x) = \dfrac{(x + 2)(x - 1) + 3}{x - 1} = x + 2 + \dfrac{3}{x - 1}$, so the answer is (A).

(b) Since $f(x) = (x + 2) + \dfrac{3}{x - 1} = (x - 1) + 3 + \dfrac{3}{x - 1} \geq 2\sqrt{(x - 1)\dfrac{3}{x - 1}} + 3$, so $f(x) \geq 2\sqrt{3} + 3$, at $x = 1 + \sqrt{3}$, by AM-GM inequality.

Given two real numbers a and b, then AM-GM inequality states that

$$\dfrac{a^2 + b^2}{2} \geq \sqrt{a^2 b^2}$$

where the equality holds if $a^2 = b^2$. Now, notice that AM-GM inequality tells us that the left-side of the inequality is greater than or equal to the right-side of the inequality. If one side of the inequality is a number, then we may safely conclude that there is a fixed minimum(or maximum) value. You must be extra-careful of when to use AM-GM inequality. Have a look at the following example. If we want to find out the minimum value of $x^2 + 1$ for real x, we use trivial inequality. At $x = 0$, $x^2 + 1 = 1$ is the minimum value. On the other hand, if we use AM-GM inequality, we get $x^2 + 1 \geq 2\sqrt{x^2 \cdot 1} = 2|x|$. The inequality turns into equality if $x^2 = 1$. At $x = \pm 1$, $x^2 + 1 = 2|x|$. Does this mean that the minimum value occurs at $x = 1$ or $x = -1$? Not really. Even though the graph of $x^2 + 1$ touches the graph of $y = 2|x|$ at $x = \pm 1$, the minimum value occurs at $x = 0$.

7

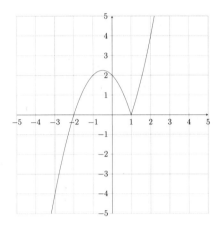

8

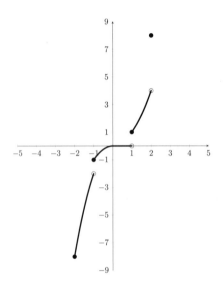

9 Since $(\sqrt{2} - 1, 10)$ passes through the line we get

$$p(\sqrt{2} - 1)^2 + \frac{q}{\sqrt{2} - 1} = 10$$
$$p(3 - 2\sqrt{2}) + q(\sqrt{2} + 1) = 10$$
$$q - 2p = 0$$
$$3p + q = 10$$

Hence, $p = 2$ and $q = 4$. Therefore, the sum of p and q is 6.

$\boxed{10}$

(a) surjective, meaning that we can always find x-value for a given y-value.
(B) injective, meaning that different x-values result in different y-values.

$\boxed{11}$ The correct answer is (D) since $g(f(1)) = g(5) = -5 + 3 = -2$.

$\boxed{12}$ Since the slope is $\dfrac{f(5) - f(-5)}{5 - (-5)} = \dfrac{1}{5}$, we get $\dfrac{f(15) - f(-15)}{15 - (-15)} = \dfrac{1}{5}$. Hence, the correct answer is (D).

$\boxed{13}$ Notice that $r^2 + s^2 = (r+s)^2 - 2rs = (-a)^2 - 2(3)$. Since a is real, $a^2 - 6 \geq -6$, by trivial inequality.

$\boxed{14}$ Recall that the graph of $y = a(x - h)^2 + k$ is symmetric about $x = h$. Since $f(3) = f(9)$, it implies that $h = \dfrac{3 + 9}{2}$. hence, $h = 6$.

$\boxed{15}$ Let r and s be the solutions of $x^2 + 3x + 4 = 0$. Then, $r + s = -3$ and $rs = 4$. The equation $x^2 - (r - 1 + s - 1)x + (r - 1)(s - 1) = x^2 - (r + s - 2)x + (rs - r - s + 1) = x^2 - (-5)x + (4 + 3 + 1) = x^2 + 5x + 8$. Hence, the answer is (C).

$\boxed{16}$ The answer is (A). Since the polynomial function is continuous, the graph is connected. $f(-1) < 0$ and $f(0) > 0$ imply that there is a zero between $x = -1$ and $x = 0$ by the Intermediate Value Theorem.

$\boxed{17}$ If $y = \dfrac{1}{20}(x + 1)^2(x - 4)^3$, the graph is (a) cutting through the x-axis at $x = -3$ and $x = 3$ and (b) tangent to the x-axis at $x = 2$. It is (c) neither increasing nor decreasing function.

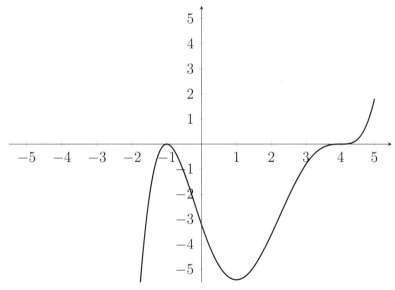

$\boxed{18}$ Let $\sqrt[3]{2 - x^3} = a$ and $\sqrt[3]{2 + x^3} = b$. Then, $a + b = 2$ and $a^3 + b^3 = 4$. Using $(a + b)^3 = a^3 + b^3 + 3ab(a + b)$, we get $8 = 4 + 3\sqrt[3]{4 - x^6}(2)$. Therefore, $4 - x^6 = \left(\dfrac{2}{3}\right)^3$, so $x^6 = 4 - \dfrac{8}{27} = \dfrac{100}{27}$.

19 We are solving for $x^2(x-1) = \dfrac{-4}{27}$. Have a look at the graph of $y = x^2(x-1)$. It has its inflection point (or point of symmetry) at $x = \dfrac{1}{3}$. At this point, $y = -\dfrac{2}{27}$. Use the symmetric property of cubic function to find out that $x = -\dfrac{1}{3}$.

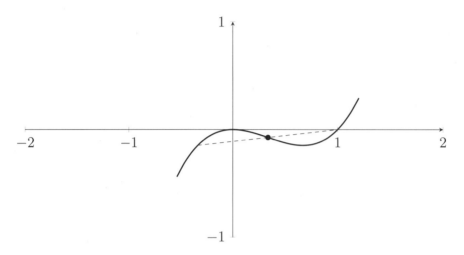

20

(a) $2x^3 - 6x^2 - 20x + 48 = 2(x-2)(x+3)(x-4) = 0$, so $x = 2$, -3 and 4.

(b) $12x^3 - 107x^2 - 15x + 54 = (x-9)(3x-2)(4x+3) = 0$, so $x = 9$, $\dfrac{2}{3}$, and $\dfrac{-3}{4}$.

21 Direct substitution tells us that

$$
\begin{aligned}
p(x) &= f(x+1) \\
&= (x+2)^3 \\
&= x^3 + 6x^2 + 12x + 8
\end{aligned}
$$

22

(a) The quotient is $x^2 + x + 7$ and the remainder is 18.

(b) Since $f(x) = (x-3)(x^2+x+7) + 18$, a value of x greater than 3 will make $f(x)$ greater than 18. Hence, it will never be 0. Therefore, $x = 3$ is an upperbound.

23 First, $p(0) = 6$ and $p(1) = -6$. Since $p(x)$ is a polynomial of x, the function $y = p(x)$ is continuous. Therefore, according to IVT, there exists at least one x between 0 and 1 such that $p(x) = 0$.

24 We cannot use AM-GM inequality, since invariants do not show up in the process of using AM-GM inequality. Hence, we need to use a different approach. We are looking for the minimum sum of the distance between $(x,0)$ and $(5,\pm3)$ and that between $(x,0)$ and $(0,\pm2)$. The shortest sum of distances from $(5,\pm3)$ to $(x,0)$ and from $(x,0)$ to $(0,\pm2)$ occurs at $x = 2$. The shortest distance, in fact, is $5\sqrt{2}$.

25

$$\sqrt{x + \sqrt{4x + \sqrt{16x + 1025}}} = 1 + \sqrt{x}$$

$$x + \sqrt{4x + \sqrt{16x + 1025}} = x + 2\sqrt{x} + 1$$

$$4x + \sqrt{16x + 1025} = 4x + 4\sqrt{x} + 1$$

$$16x + 1025 = 16x + 8\sqrt{x} + 1$$

$$8\sqrt{x} = 1024$$

$$\sqrt{x} = 128$$

$$x = 2^{14}$$

Hence, the answer is $16(= 2 + 14)$.

26 Perform caseworks. If $x \leq 2$, then $x = 5$, which contradicts the original assumption. If $2 \leq x \leq 3$, then $x = 3$. If $3 \leq x \leq 4$, then $x = 3$. If $4 < x$, then $x = 5$. Hence, there are two solutions $x = 3$ and $x = 5$. The sum must be 8.

27 Since the graph must be a rhombus with the vertices at $(\pm 5, 0)$ and $(0, \pm 3)$, we get the area of $\dfrac{10 \times 6}{2} = 30$.

28
At $(x, y) = (-2, 7)$, $y = 7$ meets the graph of $f(x)$ at three points. On the other hand, if $y = 0$, there are two points of intersection. Hence, $0 < k < 7$ satisfies four intersection points between $y = |x^2 + 4x - 3|$ and $y = k$.

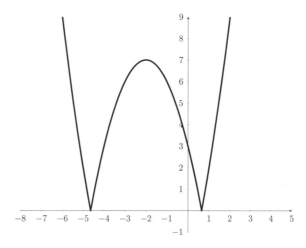

29 Since $|\dfrac{nx}{2021} - 1| < \dfrac{n}{2021}$ turns into $-1 < x - \dfrac{2021}{n} < 1$, if there is only one integer solution for x, this implies that $x - \dfrac{2021}{n} = 0$. Hence, $x = \dfrac{2021}{n}$. Since x is an integer, n must be a factor of 2021. Therefore, $n = 1, 43, 47, 2021$. There are four possible n values.

30 If $x < 8$, then $|-x + 8| = |-x + 4|$. Then $8 - x = x - 4$, so $x = 6$ if $4 < x < 8$. If $x < 4$, then there is no solution. Therefore, the only solution is $x = 6$.

31 The graph of $f(x-1)+1 = f(x)$ is identical with that of $y = \lfloor x \rfloor$.

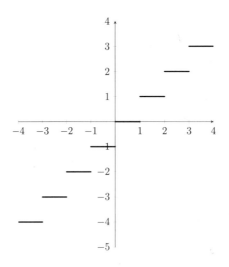

32

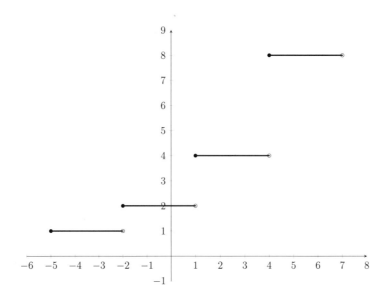

33 Let $y = \dfrac{2x^2 + 3x + 1}{x - 1}$. Then, $yx - y = 2x^2 + 3x + 1$. Hence, $2x^2 + (3-y)x + (1-y) = 0$. Since x must be a real number, we solve $(3-y)^2 - 4(2)(1-y) \geq 0$. Then, $y \leq 7 - 4\sqrt{3}$ or $7 + 4\sqrt{3} \leq y$.

34 Since $x - 5000 \geq 0$, $x - 5000 \leq 6400$ and $x - 5000 \leq 71^2 = 5041$, we get $5000 \leq x \leq 10041$. In the denominator, $x \neq 5000, 5001, 5002, \cdots, 5040$. Hence, the domain includes integers ranging from 5041 to 10041. There are 5001 integers.

35 First off, $x \neq 0$ and $x \neq \dfrac{4}{3}$. Assuming that these two values are excluded, we turn the original $g(x)$ into

$$\frac{3x-2}{\dfrac{2}{x}-\dfrac{3}{3x-4}} = \frac{3x-2}{\dfrac{6x-8-3x}{3x^2-4x}}$$
$$= \frac{x(3x-2)(3x-4)}{3x-8}$$

Hence, $x \neq \dfrac{8}{3}$. Thus, there are three values of x that should be excluded from the set of real numbers, i.e, $x = 0, \dfrac{4}{3}, \dfrac{8}{3}$.

36

(a) No, there is no such value of x for $p(x) = 3$ since the domain is restricted from 4 to 9 and $p(\dfrac{25}{9}) = 0$ where $\dfrac{25}{9} < 4$.

(b) Since $p(x) = 3\sqrt{x} - 2$ is increasing function, we simply compute its range by substituting the endpoints. Hence, $p(4) = 4$ and $p(9) = 7$ implies that $4 \leq p(x) \leq 7$ for $4 \leq x \leq 9$.

37

(a) $f(g(3)) = f(0) = 5$ and $g(f(3)) = g(11) = 6\sqrt{2}$.

(b) If $f(x) = 17$, then $2x = 12$, so $x = 6$. Hence, $x = g(a) = 6$, so $3\sqrt{a-3} = 6$. Thus, $a = 7$.

(c) $g(f(-5)) = g(-5) = 3\sqrt{-8}$ is not real, so $g(f(-5))$ does not exist.

(d) Domain of h is $[-1, \infty)$ since $g(f(x)) = 3\sqrt{2x+5-3} = 3\sqrt{2x+2}$.

38

(a) $f(g(x)) = \sqrt{x^2} = |x|$, so the domain is $\mathbb{R}$, the set of all real numbers.

(b) $g(f(x)) = (\sqrt{x})^2$ implies that the domain is the set of non-negative real numbers.

39 First, $\dfrac{x+2}{x-2}$ is undefined at $x = 2$. Also, $\dfrac{x+2}{x-2}$ cannot reach $y = 1$. Now, in the question, it is given that $-2 < \dfrac{x+2}{x-2} < 2$. Multiplying three sides of inequality by $(x-2)^2$ results in $-3(x-2)^2 < 4(x-2) < (x-2)^2$. Solving for $-3(x-2)^2 < 4(x-2)$, we get $x < \dfrac{2}{3}$ or $2 < x$. Similarly, solving for $4(x-2) < (x-2)^2$, we get $(x-2)(x-6) > 0$, so $x < 2$ or $6 < x$. Finding out the common parts, we get $(-\infty, \dfrac{2}{3}) \cup (6, \infty)$.

40 The following figure is the graph of $xf(x)$ as $x \geq 0$.

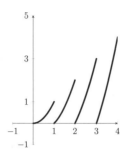

Now, taking $\{xf(x)\}$ into account, we draw the following figure.

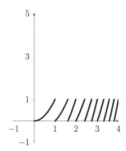

41 By sketching the graph of $\sqrt{x} - \lfloor \sqrt{x} \rfloor$, we get the following figure.

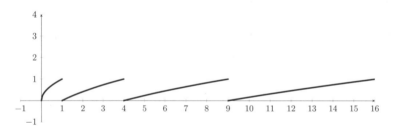

Since we are looking for intersection points between this graph and the graph of $y = \dfrac{x}{36}$, there are five solutions, in each interval whose endpoints are perfect squares of integers.

In fact, we may solve it using algebra. Let $x = r^2$ where r is a non-negative real number. Then, $r - \lfloor r \rfloor = \dfrac{r^2}{36}$. If $0 \leq r < 1$, then $r = \dfrac{r^2}{36}$, so $r = 0$. If $1 \leq r < 2$, then $r - 1 = \dfrac{r^2}{36}$, so $r = 18 - 12\sqrt{2}$. If $2 \leq r < 3$, then $r - 2 = \dfrac{r^2}{36}$, so $r = 18 - 6\sqrt{5}$. We continue this process until r reaches a negative value. In fact, $x = 0$, $(18 - 12\sqrt{2})^2$, $(18 - 6\sqrt{5})^2$, $(18 - 6\sqrt{6})^2$, $(18 - 6\sqrt{7})^2$. Hence, there are five solutions.

42 Notice that $f(x)$ and $g(x)$ are inverse functions to each other. Hence, the solutions to $f(x) = g(x)$ indicate that they satisfy $f(x) = x$ Therefore, $2x - 3 = x$ implies that $x = 3$ is also the solution to $f(x) = g(x)$.

43 Since $mx + b = \dfrac{1}{m}(x - b)$, we get $m^2 = 1$. Thus, $m = \pm 1$. The product of these two values is -1.

44 There are two sets of parabola drawn with the directrix of x-axis and y-axis and the focus at $(10, 6)$. Now, let the point of intersection be I. Then, by the definition of parabola, we know that $I(k, k)$ for some constant k. Hence, $k = \sqrt{(10-k)^2 + (6-k)^2}$ implies that $k^2 - 32k + 136 = 0$. Hence, $k = 16 \pm 2\sqrt{30}$. Since the condition stated in the question tells us that $k < 6$, we conclude that $(16 - 2\sqrt{30}, 16 - 2\sqrt{30})$ is the point of intersection.

Solutions for Topic **3**

1

(a) $f(x) = \left(\dfrac{1}{2}\right)^x - 2$ has the domain of $\mathbb{R}$ and the range of $\{y | y > -2\}$.

(b) $f(x) = \log_2(x^2 - 4x + 3)$ has the domain of $\{x | x > 3 \text{ or } x < 1\}$ and the range of $\mathbb{R}$.

2 First, $\log_{\frac{1}{9}}(x)$ implies that $x > 0$. Also, $\log_9(\log_{\frac{1}{9}}(x)) > 0$ implies that $x < \dfrac{1}{9}$. Therefore, $0 < x < \dfrac{1}{9}$.

3

(a)   (b)

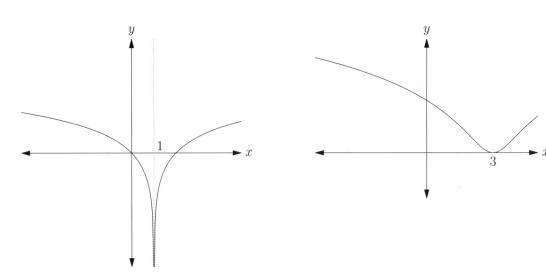

4 Since $\dfrac{k}{m} = f^{-1}(n)$, we get $f(\dfrac{k}{m}) = n$. Since $f(2k) = 2e$, we get $m = \dfrac{1}{2}$ and $n = 2e$. Hence, $mn = e$.

5 The intersection point occurs at $x = 1$. For $0 \le x \le 1$, $2^{x-1} = 0$. For $x = 1$, $2^{x-1} = 1$. Hence, $x = 1$ is the only solution.

6

$$\log_9(x+3)^2 + \log_9(x-2)^2 = \log_{27}(2(x+3))^3$$
$$\log_9((x+3)(x-2))^2 = \log_{27}(2(x+3))^3$$
$$\log_3 |(x+3)(x-2)| = \log_3(2(x+3))$$

Hence, $x = 4$ or $x = 0$. The sum of solutions is 4.

$$\boxed{7}$$

$$\log_2 \frac{2}{1} + \log_2 \frac{3}{2} + \cdots + \log_2 \frac{2019}{2018} + \log_2 \frac{2020}{2019} = \log_2(2020)$$

Hence, $\lfloor \log_2(2020) \rfloor = 10$, since $\log_2(2048) = 11$.

$$\boxed{8}$$

$$\begin{aligned}
\log_2 \left(\frac{64^x}{16^y} \right) &= \log_2(64^x) - \log_2(16^y) \\
&= x \log_2(64) - y \log_2(16) \\
&= 6x - 4y
\end{aligned}$$

Hence, $a = 6$ and $b = -4$.

$\boxed{9}$ Since $\log_{10}(2^{2019}) = 2019 \log_{10}((2) = 607.719$, we get 608 digits used in 2^{2019}.

$\boxed{10}$ Since $\log_a(b) + \log_b(a) \geq 2\sqrt{\log_a(b) \cdot \log_b(a)} = 2$, we get

$$\begin{aligned}
\log_a b + \log_b a &\geq 2 \\
-(\log_a b + \log_b a) &\leq -2 \\
2 - (\log_a b + \log_b a) &\leq 2 - 2 \\
2 - (\log_a b + \log_b a) &\leq 0
\end{aligned}$$

Hence, the maximum value is 0.

$$\boxed{11}$$

$$\begin{aligned}
\log_y(x) + \log_x(y) &= \frac{17}{4} \\
\frac{1}{\log_x(y)} + \log_x(y) &= \frac{17}{4}
\end{aligned}$$

It is easy to check that that $\log_y(x) = 4$. Hence, $y = 2$ and $x = 16$. The value of x is 16.

$$\boxed{12}$$

$$\begin{aligned}
\log_{10}(\sqrt{10} \cdot x^{\log_{10}(x)}) &= \log_{10}(x\sqrt{x}) \\
\frac{1}{2} + (\log_{10}(x))^2 &= \frac{3}{2} \log_{10}(x) \\
2(\log_{10}(x))^2 - 3(\log_{10}(x)) + 1 &= 0
\end{aligned}$$

Hence, $\log_{10}(x) = \frac{1}{2}, 1$. Thus, $x = \sqrt{10}, 10$. The product of the positive roots of the given equation is $10\sqrt{10}$.

$\boxed{13}$ Since $\log_b(c) = \frac{20}{19}$, we get $b^{\frac{20}{19}} = c$. Hence, $bc = c^{\frac{19}{20}} \cdot c = c^{\frac{39}{20}}$. Hence, $p + q = 59$.

Solutions for Topic 4

1 First, $\sin(30°) = \dfrac{1}{2}$, $\cos(30°) = \dfrac{\sqrt{3}}{2}$ and $\tan(30°) = \dfrac{\sqrt{3}}{3}$. Second, $\sin(45°) = \dfrac{\sqrt{2}}{2}$, $\cos(45°) = \dfrac{\sqrt{2}}{2}$ and $\tan(45°) = 1$. Third, $\sin(60°) = \dfrac{\sqrt{3}}{2}$, $\cos(60°) = \dfrac{1}{2}$, and $\tan(60°) = \sqrt{3}$.

2

(a) $\cos(\theta) = \sin(\theta) = \dfrac{\sqrt{2}}{2}$.

(b) $\cos(\phi) = \dfrac{\sqrt{10}}{10}$ and $\sin(\phi) = \dfrac{3\sqrt{10}}{10}$.

(c) $\sin(\theta) = \dfrac{\sqrt{3}}{2}$ and $\tan(\theta) = \sqrt{3}$

(d) $\cos(\theta) = \sqrt{1-x^2}$ and $\tan(\theta) = \dfrac{x}{\sqrt{1-x^2}}$

3

(a) $\theta = 45°$ since $0° \le \theta \le 90°$

(b) $\theta = 45° + 90° = 135°$ since $90° \le \theta \le 180°$

(c) $\theta = 45° + 180° = 225°$ since $180° \le \theta \le 270°$

(d) $\theta = 45° + 270° = 315°$ since $270° \le \theta \le 360°$

4

(a)

$$\frac{\cos(135°)}{\sin(120°) + \cos(150°)} = \frac{-\cos(45°)}{\sin(60°) - \cos(30°)}$$
$$= \text{undefined}$$

(b)

$$\tan(135°) - \tan(225°)\tan(-45°) = -\tan(45°) - \tan(45°)(-\tan(45°))$$
$$= -1 - (1)(-1)$$
$$= 0$$

(c)

$$\frac{\sin(30°)}{\cos(60°)} = \frac{\sin(30°)}{\sin(30°)}$$
$$= 1$$

(d)

$$\sec(135°) = \frac{1}{\cos(135°)}$$
$$= \frac{1}{-\cos(45°)}$$
$$= \frac{1}{-\sqrt{2}/2}$$
$$= -\sqrt{2}$$

(e)

$$\tan(210°) = \tan(30°)$$
$$= \frac{\sqrt{3}}{3}$$

(f)

$$\cos(330°) + \sin(300°) = \cos(30°) - \sin(60°)$$
$$= 0$$

5

(a)

$$\cos(90° - \theta)\sin(\theta) = \sin^2(\theta)$$
$$= \frac{1}{4}$$

(b)

$$\cot(\theta)\tan(90° - \theta) = \cos^2(\theta)$$
$$= \frac{1}{\tan^2(\theta)}$$
$$= \frac{1}{(1/\sqrt{3})^2}$$
$$= 3$$

(c)

$$\cos^2(\theta) + \sin^2(\theta) = 1$$

(d)

$$1 + \tan^2(\theta) = \sec^2(\theta)$$
$$= \frac{1}{\cos^2(\theta)}$$
$$= \frac{1}{(\sqrt{3}/2)^2}$$
$$= \frac{4}{3}$$

(e)

$$1 + \cot^2(\theta) = \csc^2(\theta)$$
$$= \frac{1}{\sin^2(\theta)}$$
$$= \frac{1}{(1/2)^2}$$
$$= 4$$

(f)

$$\sec^2(90° - \theta) - \tan^2(90° - \theta) = \frac{1}{\cos^2(90° - \theta)} - \frac{\sin^2(90° - \theta)}{\cos^2(90° - \theta)}$$
$$= \frac{\cos^2(90° - \theta)}{\cos^2(90° - \theta)}$$
$$= 1$$

6

(a) Since $\sin(\theta) < 0$, we get $\sin(\theta) = -\dfrac{\sqrt{3}}{2}$ and $\tan(\theta) = -\sqrt{3}$. Hence, $\sin(\theta) + \tan(\theta) = -\dfrac{3\sqrt{3}}{2}$.

(b) Since $\sin(\theta) < 0$, we get $\sin(\theta) = -\dfrac{\sqrt{5}}{3}$ and $\tan(\theta) = \dfrac{\sqrt{5}}{2}$.

(c) Since $\cos(\theta) < 0$, we get $\cot(\theta) = \dfrac{3}{4}$ and $\sec(\theta) = -\dfrac{5}{3}$.

(d) Since $\sin(\theta) : \cos(\theta) = 1 : 3$, let $\sin(\theta) = k$ and $\cos(\theta) = 3k$ where $k < 0$. Hence, $10k^2 = 1$ to get $k = -\dfrac{\sqrt{10}}{10}$. Therefore, $\sec(\theta) = -\dfrac{\sqrt{10}}{3}$ and $\csc(\theta) = -\sqrt{10}$.

7 Let $\sin A = 7k$ and $\cos A = 4k$. Hence, $65k^2 = 1$ implies $k = \dfrac{\sqrt{65}}{65}$. Therefore, $\sin A = \dfrac{7\sqrt{65}}{65}$.

8 Since $789 = 2(360) + 69$, we get $n = 69$.

9 Since $n + 543 = 180, 540, 900, \cdots$, we get $n = -3$ which is between $-90 \le n \le 90$.

Solutions for Topic 5

1 Cross out $f(2021)$ by $f(1009)$ until $f(1013)$ is crossed out by $f(1)$. Now, there is only one term left in the numerator $f(1011)$, which is equal to π. Hence, the answer is π.

2 It is well-known that $\dfrac{100}{\pi} \approx 31.83$. Hence, the answer is (B).

3 First, $\sin(k\pi) = 0$ for integer k. Second, $0.001 < x < 0.01$ implies that $\dfrac{1}{1000} < \dfrac{1}{k\pi} < \dfrac{1}{100}$. Hence, $\dfrac{100}{\pi} < k < \dfrac{1000}{\pi}$. Therefore, $k = 32, 33, 34, \cdots, 318$. There are 287 integer value of k satisfying $f(g(x)) = 0$. Hence, the answer must be 287.

4 For $f(f(x)) = 0$, we get $2\pi f(x) = k\pi$ for some integer k. Hence, $f(x) = \pm\dfrac{1}{2}$, 0, ± 1. Solving for $f(x) = \pm\dfrac{1}{2}$, we get 16 solutions. For $f(x) = 0$, we get 9 solutions. For $f(x) = \pm 1$, we get 8 solutions. Hence, there are 33 solutions in total.

5 Since $\cos(\dfrac{1}{x}) = 1$ at $\dfrac{1}{x} = 2k\pi$ for some integer k, we get

$$\frac{1}{1000} < x < \frac{1}{100}$$
$$\frac{1}{1000} < \frac{1}{2k\pi} < \frac{1}{100}$$
$$\frac{100}{\pi} < 2k < \frac{1000}{\pi}$$
$$31.83 < 2k < 318.3$$

Hence, $k = 16, 17, \cdots, 159$. Therefore, there are 144 real solutions.

6 Sketch the graph directly to find out that $f(f(x))$ has V-shaped portion appearing twice in $[0, 1]$ and $[1, 2]$. Hence, there are 8 intersection points between $y = f(f(x))$ and $y = \dfrac{1}{2}$.

7 Since $-1 \leq \cos(\pi x) \leq 1$ for $x \in [0, 1]$, let $k = \cos(\pi x)$. Then, $\sin(k) = 1/2$ implies that $k = \dfrac{\pi}{6}$ or $\dfrac{5\pi}{6}$. There is only one solution $k = \dfrac{\pi}{6}$ between -1 and 1. Hence, the number of real x's is 1.

8 Since $\tan(x)$ is increasing for $0 \leq x \leq \dfrac{\pi}{4}$, we get

$$\tan(0) \leq \tan(x) \leq \tan(\frac{\pi}{4})$$
$$3\tan(0) \leq 3\tan(x) \leq 3\tan(\frac{\pi}{4})$$
$$0 \leq 3\tan(x) \leq 3$$
$$2 + 0 \leq 2 + 3\tan(x) \leq 2 + 3$$

Hence, $2 + 3\tan(x)$ is between 2 and 5, inclusive. The range must be $[2, 5]$.

(a)

(b)

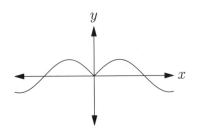
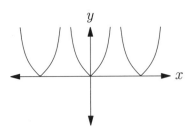

10

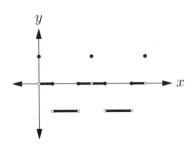

11 Period of accumulative functions $f \pm g \pm h$ is the least common multiple of individual periods. Hence, we get the period of 4 for $\sin(\frac{\pi}{2}x)$, 6 for $\cos(\frac{\pi}{3}x)$ and 1 for $\tan(\pi x)$. Hence, the least common multiple of 4, 6 and 1 is 12. The answer is (E).

12 Since $f(x)$ has the period of p, $f(p) = f(0)$. Hence,
$f(0) = \sqrt{1 + \sin(0)} + \sqrt{1 - \sin(0)} = 2$. The answer is (D).

13

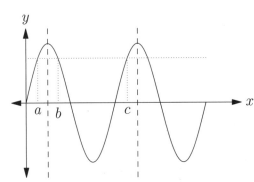

Using the line symmetry at $x = \dfrac{\pi}{2}$ and $x = \dfrac{5\pi}{2}$, we get

$$a = \sin^{-1}(\frac{3}{4})$$

$$b = \pi - \sin^{-1}(\frac{3}{4})$$

$$c = 2\pi + \sin^{-1}(\frac{3}{4})$$

Hence, $a + b + c = 3\pi + \sin^{-1}(\frac{3}{4})$. Also, $f(a+b+c) = f(a+b+c-2\pi)$ by periodicity. Therefore, we get -3 by line symmetry. The answer is (D).

$\boxed{14}$ Notice that $f(\pi - x) = \begin{cases} 1 & (x \leq \pi) \\ 0 & (x > \pi) \end{cases}$. Therefore,

$f(x)f(\pi - x) = \begin{cases} 0 & (x > \pi) \\ 1 & (0 \leq x \leq \pi) \\ 0 & (x < 0) \end{cases}$. Thus, there are two intersections between

$g(x) = f(x)f(\pi - x)\sin(2x)$ and $h(x) = \frac{1}{3}$.

$\boxed{15}$ As one can see from the figure, $A = 2$. Its period must be double the length between $-\frac{3\pi}{2}$ and $\frac{3\pi}{2}$, which is 6π. Thus, $\frac{2\pi}{B} = 6\pi$, so $B = \frac{1}{3}$. The product of A and B is $\frac{2}{3}$.

$\boxed{16}$ The graph of $|\cot(2x)|$ looks similar to $|\tan(2x)|$, simply translated by $\frac{\pi}{2}$. At $x = \frac{\pi}{4}$, the graph is above $y = 2|x - \frac{\pi}{4}|$. Hence, the maximum value of k is 2.

$\boxed{17}$

(a) Let $\cos(x) = t$. Then, for all real x, we get $-1 \leq t \leq 1$. Hence, $y = |\cos(x) - 2| + 2$ is between 3 and 5, inclusive.

(b) Let $\sin(x) = t$. Then for all real x, we get $-1 \leq t \leq 1$. Hence, $y = 3 - |2 - \sin(x)|$ is between 0 and 2, inclusive.

$\boxed{18}$

(a) Let $\cos(x) = t$. Then for $0 \leq x \leq \pi$, we get $-1 \leq t \leq 1$. Therefore, $y = (t - \frac{5}{2})^2 - \frac{9}{4}$. Hence, maximum value is 10 and minimum value is 0.

(b) Let $\tan(x) = t$. Then for $-\frac{\pi}{4} \leq x \leq \frac{\pi}{4}$, we get $-1 \leq t \leq 1$. Hence, $t^2 + t = (t + \frac{1}{2})^2 - \frac{1}{4}$ implies that the maximum value is 2 and minimum value is $-\frac{1}{4}$.

$\boxed{19}$ Assume $a > 0$. Then, we will perform caseworks on a. If $0 < a < 1$, then $-\cos^2(x) + 2a\cos(x) = -(\cos(x) - a)^2 + a^2$ has its maximum at (a, a^2). Since the maximum value is given by 3, it contradicts the given assumption that $a < 1$. If $1 \leq a$, then the maximum value occurs at $\cos(x) = 1$. Hence, $-1 + 2a = 3$. Therefore, $2a = 4$, so $a = 2$.

$\boxed{20}$ Let $\sin(x) = k$. Then, $y = \frac{2k}{k+4} = 2 - \frac{8}{k+4}$ has its maximum at $k = 1$ and minimum at $k = -1$. At $k = 1$, we get $\frac{2}{5}$. At $k = -1$, we get $-\frac{2}{3}$. Thus, the maximum value is $\frac{2}{5}$ and minimum value is $-\frac{2}{3}$. The difference must be $\frac{16}{15}$.

21

(a) Let $\theta = 2x$ Then, $0 \leq \theta \leq 4\pi$. Since $\cos(\theta) = \sin(\theta)$, we get $\theta = \dfrac{\pi}{4}, \dfrac{5\pi}{4}, \dfrac{9\pi}{4}$ and $\dfrac{13\pi}{4}$.

Hence, $x = \dfrac{\pi}{8}, \dfrac{5\pi}{8}, \dfrac{9\pi}{8}$ and $\dfrac{13\pi}{8}$.

(b) Multiply both sides by $\sin^2(x)$. Then, $\cos^2(x) + 3\sin(x) = 3\sin^2(x)$. Hence, $(1 - \sin^2(x)) + 3\sin(x) = 3\sin^2(x)$ is solved by

$$4\sin^2(x) - 3\sin(x) - 1 = (4\sin(x) + 1)(\sin(x) - 1)$$

Hence, $\sin(x) = 1$ or $\sin(x) = -\dfrac{1}{4}$. Since x is in the first or second quadrants, we get $\sin(x) = 1$. Therefore, $x = 90°$.

22 Let $k = x + \dfrac{\pi}{4}$. Then, $\dfrac{\pi}{4} \leq k \leq \dfrac{9\pi}{4}$. Solving $\sin(\theta) = -\dfrac{1}{2}$, we get $\theta = \dfrac{7\pi}{6}, \dfrac{11\pi}{6}$. Hence, $x = \dfrac{11\pi}{12}, \dfrac{19\pi}{12}$.

23

$$2\cos(x)\cot(x) + 1 = \cot(x) + 2\cos(x)$$
$$2\cos(x)\cot(x) + 1 - \cot(x) - 2\cos(x) = 0$$
$$(\cot(x) - 1)(2\cos(x) - 1) = 0$$

Hence, $\cot(x) = 1$ or $\cos(x) = \dfrac{1}{2}$. Therefore, $x = \dfrac{\pi}{4}$ or $\dfrac{\pi}{3}$.

24 Let $\sin^{-1}(\sin(\dfrac{2\pi}{3})) = \theta$. Then θ is defined on $[-\dfrac{\pi}{2}, \dfrac{\pi}{2}]$. Hence, $\sin(\dfrac{2\pi}{3}) = \sin(\theta)$ for such defined interval of θ. We get $\theta = \dfrac{\pi}{3}$.

25 This is *undefined*, since the domain of $\sin^{-1}(x)$ is $[-1, 1]$.

26

(a) $\arcsin(0) = 0$ (b) $\cos^{-1}(\dfrac{\sqrt{3}}{2}) = \dfrac{\pi}{6}$ (c) $\arctan(1) = \dfrac{\pi}{4}$

27

(a) $\arcsin(\sin(39\pi/5)) = -\pi/5$ (b) $\arcsin(\sin(37\pi/5)) = -2\pi/5$

(c) $\sin(\arccos(0.6)) = 0.8$ (d) $\sin(\arccos(-0.6)) = 0.8$

28

(a) $\cot^{-1}(1) = \dfrac{\pi}{4}$ (b) $\cot^{-1}(-1) = \dfrac{3\pi}{4}$

29

(a) The range of inverse trigonometric functions is radians and the domain is the range of trigonometric functions. With identical trigonometric function range with two different trigonometric functions, i.e., $\sin(x)$ and $\cos(x)$, the sum of angles add upto $\dfrac{\pi}{2}$.

(b) First, $\arccos(\cos(7)) = \theta$ implies that $\cos(7) = \cos(\theta)$ where $0 \leq \theta \leq \pi$. Since 7 is a bit greater than 2π, we get $7 - 2\pi = \theta$. Similarly, $\arcsin(\sin(5)) = \theta$ implies that $\sin(5) = \sin(\theta)$, where $-\pi/2 \leq \theta \leq \pi/2$. Since 5 is between $3\pi/2$ and 2π, we get $\theta = 5 - 2\pi$. Hence, $\arccos(\cos(7)) + \arcsin(\sin(5)) = 12 - 4\pi$.

30 For $-\dfrac{\pi}{2} \leq x \leq \dfrac{\pi}{2}$, then $y + x = \dfrac{\pi}{2}$. For $-\dfrac{3\pi}{2} \leq x \leq \dfrac{\pi}{2}$, then $y - x = -\dfrac{3\pi}{2}$. Hence, the area of the triangle bounded by the region has the area of

$$\frac{1}{2} \times 2\pi \times \pi = \pi^2$$

Solutions for Topic 6

1

(a)

$$\frac{1}{1-\cos(\theta)} + \frac{1}{1+\cos(\theta)} = \frac{2}{1-\cos^2(\theta)}$$
$$= \frac{2}{\sin^2(\theta)}$$
$$= 2\csc^2(\theta)$$

(b)

$$\frac{1}{1-\sin(x)} - \frac{1}{1+\sin(x)} = \frac{2\sin(x)}{1-\sin^2(x)}$$
$$= \frac{2\sin(x)}{\cos^2(x)}$$
$$= 2 \cdot \frac{\sin(x)}{\cos(x)} \cdot \frac{1}{\cos(x)}$$
$$= 2\tan(x)\sec(x)$$

(c)

$$(1+\sec\theta)(1-\cos\theta) = 1 + \sec\theta - \cos\theta - 1$$
$$= \frac{1}{\cos\theta} - \cos\theta$$
$$= \frac{1-\cos^2\theta}{\cos\theta}$$
$$= \frac{\sin^2\theta}{\cos\theta}$$
$$= \sin\theta \times \frac{\sin\theta}{\cos\theta}$$
$$= \sin\theta\tan\theta$$

(d)

$$\frac{1}{\tan\theta + \cot\theta} = \frac{\sin\theta\cos\theta}{\sin^2\theta + \cos^2\theta}$$
$$= \frac{\sin\theta\cos\theta}{1}$$
$$= \sin\theta\cos\theta$$

2

(a)

$$\cos(75°) = \cos(30° + 45°)$$
$$= \cos(30°)\cos(45°) - \sin(30°)\sin(45°)$$
$$= \frac{\sqrt{3}}{2} \cdot \frac{\sqrt{3}}{2} - \frac{1}{2} \cdot \frac{\sqrt{3}}{2}$$
$$= \frac{\sqrt{6} - \sqrt{2}}{4}$$

(b)

$$\sin(15°) = \sin(45° - 30°)$$
$$= \sin(45°)\cos(30°) - \cos(45°)\sin(30°)$$
$$= \frac{\sqrt{2}}{2} \cdot \frac{\sqrt{2}}{2} - \frac{\sqrt{2}}{2} \cdot \frac{1}{2}$$
$$= \frac{\sqrt{6} - \sqrt{2}}{4}$$

$\boxed{3}$

$$\tan(2x) = \frac{\sin(2x)}{\cos(2x)}$$
$$= \frac{2\sin(x)\cos(x)}{\cos^2(x) - \sin^2(x)}$$
$$= \frac{2\sin(x)/\cos(x)}{1 - \sin^2(x)/\cos^2(x)}$$
$$= \frac{2\tan(x)}{1 - \tan^2(x)}$$

$\boxed{4}$ Since $\tan(3x) = \dfrac{\tan(2x) + \tan(x)}{1 - \tan(2x)\tan(x)}$, we use $\tan(x) = 3$ and

$\tan(2x) = \dfrac{2\tan(x)}{1 - \tan^2(x)} = \dfrac{6}{1-9} = -\dfrac{3}{4}$ to get $\tan(3x) = \dfrac{-3/4 + 3}{1 - (-3/4)(3)} = \dfrac{9/4}{13/4} = \dfrac{9}{13}$.

$\boxed{5}$

$$\sin 6° \cos 12° \cos 24° \cos 48° = \frac{2\cos 6° \sin 6° \cos 12° \cos 24° \cos 48°}{2\cos 6°}$$
$$= \frac{\sin 12° \cos 12° \cos 24° \cos 48°}{2\cos 6°}$$
$$= \frac{2\sin 12° \cos 12° \cos 24° \cos 48°}{4\cos 6°}$$
$$= \frac{\sin 24° \cos 24° \cos 48°}{4\cos 6°}$$
$$= \frac{2\sin 24° \cos 24° \cos 48°}{8\cos 6°}$$
$$= \frac{\sin 48° \cos 48°}{8\cos 6°}$$
$$= \frac{2\sin 48° \cos 48°}{16\cos 6°}$$
$$= \frac{\sin 96°}{16\cos(-6)°}$$
$$= \frac{1}{16}$$

$\boxed{6}$

#1. Let $\cos(36°) = k$.
#2. Then, $\cos(72°) = 2k^2 - 1$.
#3. Then, $\cos(144°) = 2(2k^2 - 1)^2 - 1$.
#4. We know that $\cos(144°) = -\cos(36°)$.
#5. Therefore, $8k^3 - 8k^2 + 1 = 0$.
#6. Then, $(2k - 1)(4k^2 - 2k - 1) = 0$.
#7. Hence, $k = (1 + \sqrt{5})/4$.
#8. The answer is $10 (= 1 + 5 + 4)$.

7

(a)
$$\cos(3x)\sin(5x) = \frac{1}{2}(\sin(8x) + \sin(2x))$$

(b)
$$\sin(5x)\sin(2x) = \frac{1}{2}(\cos(3x) - \cos(7x))$$

8

$$
\begin{aligned}
\cos(10°)\sin(20°)\sin(40°) &= \frac{1}{2}(\sin(30°) + \sin(10°))\sin(40°) \\
&= \frac{1}{2}(\frac{1}{2} + \sin(10°))\sin(40°) \\
&= \frac{1}{4}\sin(40°) + \frac{1}{2}\sin(10°)\sin(40°) \\
&= \frac{1}{4}\sin(40°) + \frac{1}{4}(\cos(30°) - \cos(50°)) \\
&= \frac{1}{4}\sin(40°) + \frac{\sqrt{3}}{8} - \frac{1}{4}\cos(50°) \\
&= \frac{\sqrt{3}}{8}
\end{aligned}
$$

9

(a)
$$\sin(3x) + \sin(2x) = 2\sin(2x)\cos(x)$$

Hence, $\sin(2x) = 0$ or $\cos(x) = 0$. Therefore, $x = 0, \frac{\pi}{2}, \pi, \frac{3\pi}{2}$, or 2π.

(b)
$$
\begin{aligned}
\cos(x) + \cos(3x) &= 2\cos(2x)\cos(x) \\
&= 2(2\cos^2(x) - 1)\cos(x)
\end{aligned}
$$

Hence, $\cos(x) = \pm\frac{\sqrt{2}}{2}$ or $\cos(x) = 0$. Hence, $x = \frac{\pi}{4}, \frac{3\pi}{4}, \frac{5\pi}{4}, \frac{7\pi}{4}$, or $\frac{\pi}{2}$ or $\frac{3\pi}{2}$.

10 $\sin(6x) + \sin(4x) = 2\sin(5x)\cos(x)$.

11

$$
\begin{aligned}
\cos(\theta) - \sin(\theta) &= \sqrt{2}(\frac{\sqrt{2}}{2}\cos(\theta) - \frac{\sqrt{2}}{2}\sin(\theta)) \\
&= \sqrt{2}(\sin(45°)\cos(\theta) - \cos(45°)\sin(\theta)) \\
&= \sqrt{2}\sin(45° - \theta) \\
&= \sqrt{2}\sin(20°)
\end{aligned}
$$

Hence, $\theta = 25°$.

12 Let $\theta = \dfrac{\pi}{14}$. Then,

$$
\begin{aligned}
\frac{1 - 4\cos(4\theta)\sin(\theta)}{\sin(\theta)} &= \frac{2\cos(\theta) - 8\sin(\theta)\cos(\theta)\cos(4\theta)}{2\cos(\theta)\sin(\theta)} \\
&= \frac{2\cos(\theta) - 4\sin(2\theta)\cos(4\theta)}{\sin(2\theta)} \\
&= \frac{4\cos(\theta)\cos(2\theta) - 8\cos(2\theta)\sin(2\theta)\cos(4\theta)}{2\cos(2\theta)\sin(2\theta)} \\
&= \frac{4\cos(\theta)(2\cos^2(\theta) - 1) - 4\sin(4\theta)\cos(4\theta)}{\sin(4\theta)} \\
&= \frac{8\cos^3(\theta) - 4\cos(\theta) - 2\sin(8\theta)}{\sin(4\theta)} \\
&= \frac{8\cos^3(\theta) - 4\cos(\theta) - 2\cos(-\theta)}{\cos(3\theta)} \\
&= \frac{8\cos^3(\theta) - 4\cos(\theta) - 2\cos(\theta)}{\cos(3\theta)} \\
&= \frac{8\cos^3(\theta) - 6\cos(\theta)}{4\cos^3(\theta) - 3\cos(\theta)} \\
&= 2
\end{aligned}
$$

Solutions for Topic 7

1

#1. Find the semiperimeter $s = \dfrac{13 + 14 + 15}{2} = 21$.

#2. Find the area $A = \sqrt{21 \cdot 8 \cdot 7 \cdot 6} = 84$.

2

#1. Let $\theta = \angle BAC$.

#2. By the second law of cosines, $7^2 = 5^2 + 8^2 - 2 \cdot 5 \cdot 8 \cdot \cos(\theta)$.

#3. Solving $\cos(\theta) = 1/2$, we get $\theta = 60°$.

3

#1. Let $ABCD$ has its side length 1.

#2. Then, $DM = DN = \sqrt{5}/2$ and $MN = \sqrt{2}/2$.

#3. By the second law of cosines, $1/2 = 5/4 + 5/4 - 2 \cdot 5/4 \cdot \cos(\theta)$.

#4. Hence, $\cos(\theta) = 4/5$.

4

#1. Notice that $BP = AP = 5/2$ and $AB = 3$.

#2. By the second law of cosines, $9 = 25/4 + 25/4 - 2 \cdot 25/4 \cdot \cos(\theta)$, where $\theta = \angle APB$.

#3. Hence, $\cos(\theta) = 7/25$. Therefore, $\sin(\angle APB) = 24/25$.

5

#1. The triangle is isosceles, so let M be the midpoint of $\overline{AB}$.

#2. $AM = \sqrt{42}$ and $MC = \sqrt{7}$.

#3. Hence, $DM = \sqrt{2}$.

#4. Thus, $AD = \sqrt{42} \pm \sqrt{2}$.

6

#1. Apply Apollonius' Theorem to get $5^2 + 4^2 = 2(x^2 + 3^2)$.

#2. Hence, $x = \sqrt{23/2}$, so $x = \sqrt{46}/2$.

$\boxed{7}$

(a)

$$a\cos(A) = b\cos(B)$$
$$a\frac{b^2+c^2-a^2}{2bc} = b\frac{a^2+c^2-b^2}{2ac}$$
$$\frac{a}{b}(b^2+c^2-a^2) = \frac{b}{a}(a^2+c^2-b^2)$$
$$a^2(b^2+c^2-a^2) = b^2(a^2+c^2-b^2)$$
$$a^2c^2-a^4 = b^2c^2-b^4$$
$$c^2(a^2-b^2)-(a^2-b^2)(a^2+b^2) = 0$$
$$(c^2-a^2-b^2)(a^2-b^2) = 0$$

Hence, $a = b$ or $c^2 = a^2+b^2$. Therefore, a triangle ABC can be either right or isosceles.

(b)

$$\sin(A)+\sin(B) = \sin(C)(\cos(A)+\cos(B))$$
$$\frac{a}{2R}+\frac{b}{2R} = \frac{c}{2R}\left(\frac{b^2+c^2-a^2}{2bc}+\frac{a^2+c^2-b^2}{2ac}\right)$$
$$a+b = \frac{b^2+c^2-a^2}{2b}+\frac{a^2+c^2-b^2}{2a}$$
$$2ab(a+b) = a(b^2+c^2-a^2)+b(a^2+c^2-b^2)$$
$$2ab(a+b) = ab(a+b)+c^2(a+b)-(a+b)(a^2-ab+b^2)$$
$$(a+b)(a^2+b^2-c^2) = 0$$

Hence, $a^2+b^2 = c^2$, implying that a triangle ABC is right.

$\boxed{8}$ By law of sines, $\dfrac{5\sqrt{3}}{15} = \dfrac{\sin(30°)}{\sin(C)}$. Hence, $\sin(C) = \dfrac{\sqrt{3}}{2}$. Thus, $C = 60°$ or $120°$.

$\boxed{9}$ The orthic triangle has its side $a\cos(A)$ and $b\cos(B)$ with the common angle of $\pi-2C$. Hence,

$$\frac{1}{2}\cdot a\cos(A)\cdot b\cos(B)\cdot\sin(\pi-2C) = \frac{1}{2}ab\cos(A)\cos(B)\sin(2C)$$
$$= \frac{1}{4}\cdot ab\cdot\sin(C)\cdot\cos(C)$$
$$= \frac{1}{2}ab\sin(C)\cdot\frac{1}{2}\cos(C)$$
$$= 1\cdot\frac{1}{24}$$

$\boxed{10}$

#1. Let x be its width and y be its length.
#2. $x + y = 22$ and $4x = 7y$ imply that $x = 14$ and $y = 8$.
#3. Multiplying altitudes with one of the sides, we get $28 = 1/2 \cdot 14 \cdot 8 \cdot \sin(\theta)$. Hence, $\sin(\theta) = 1/2$.
#4. Therefore, $\theta = 30°$, or $150°$. In radians, they are $\pi/6$, or $5\pi/6$.

$\boxed{11}$ $||\mathbf{v}|| = \sqrt{(4)^2 + (-3)^2} = \sqrt{25} = 5$.

$\boxed{12}$ $\mathbf{u} \cdot \mathbf{v} = 1 \cdot 3 + 3 \cdot (-2) = -3$.

$\boxed{13}$ $\cos(\theta) = \dfrac{\mathbf{u} \cdot \mathbf{v}}{||\mathbf{u}||||\mathbf{v}||} = \dfrac{1}{2}$, so $\theta = 60°$.

$\boxed{14}$ $< \dfrac{26}{5}, \dfrac{13}{5} >$.

$\boxed{15}$ $8\mathbf{a} \cdot \mathbf{b} + 2\mathbf{a} \cdot \mathbf{a} - 4\mathbf{b}\mathbf{b} - \mathbf{b} \cdot \mathbf{a} = 8(0) + 2(25) - 4(9) - 0 = 14$.

$\boxed{16}$ Solving $1 \cdot x + 4 \cdot (-2) = 0$, we get $x = 8$.

$\boxed{17}$ Since $(\mathbf{u} + \mathbf{v} + \mathbf{w}) \cdot (\mathbf{u} + \mathbf{v} + \mathbf{w}) = 0$, we get
$\mathbf{u} \cdot \mathbf{u} + \mathbf{v} \cdot \mathbf{v} + \mathbf{w} \cdot \mathbf{w} + 2(\mathbf{u} \cdot \mathbf{v} + \mathbf{u} \cdot \mathbf{w} + \mathbf{v} \cdot \mathbf{w}) = 0$. Hence, $\mathbf{u} \cdot \mathbf{v} + \mathbf{u} \cdot \mathbf{w} + \mathbf{v} \cdot \mathbf{w} = -14/2 = -7$.

$\boxed{18}$

#1. First, $\overrightarrow{OQ} = r\mathbf{b} + (1 - r)\mathbf{a}$.
#2. Second, $\overrightarrow{OQ} = s\overrightarrow{OP} = s(\mathbf{b} + 3(\mathbf{a} + \mathbf{b}))$.
#3. Hence, $r = 4s$ and $1 - r = 3s$. Therefore, $s = 1/7$ and $r = 4/7$.

$\boxed{19}$

#1. $\overrightarrow{OA} = \mathbf{u} + \mathbf{v}$.
#2. $\overrightarrow{OB} = 3\mathbf{u} - 2\mathbf{v}$.
#3. $\overrightarrow{OC} = 6\mathbf{u} + m\mathbf{v}$.
#4. There exists some k such that $k\overrightarrow{AB} = \overrightarrow{AC}$.
#5. $k(2\mathbf{u} - 3\mathbf{v}) = 5\mathbf{u} + (m - 1)\mathbf{v}$.
#6. $m = -6.5$.

$\boxed{20}$

#1. $\overrightarrow{OA} = < 4, 3 >$ and $\overrightarrow{OB} = < 1, t >$.
#2. $\cos(\angle AOB) = (\overrightarrow{OA} \cdot \overrightarrow{OB})/(||\overrightarrow{OA}||||\overrightarrow{OB}||)$.
#3. $(4 + 3t)/(5\sqrt{1 + t^2}) = 2/\sqrt{5}$ implies that $11t^2 - 24t + 4 = 0$. Hence, $t = 2/11$ or 2.

$\boxed{21}$ Since $a^2 + b^2 = 25$ and $8a - 6b = 0$, then $a = 3k$ and $b = 4k$. Therefore, $a = \pm 3$ and $b = \pm 4$. Thus, $|a| = 3$ and $|b| = 4$.

22 Notice that $\overrightarrow{AP}$ is the opposite of $\overrightarrow{BC}$. Hence, $\overrightarrow{AP}$ and $\overrightarrow{BC}$ are parallel, and $AP = BC$. Therefore, $\angle CAP = 120°$. Hence, $PC = 2\sqrt{3}$.

23

#1. First, $|\overrightarrow{PA} + \overrightarrow{PB}|$ passes through $(3, 2)$, the midpoint of A and B.
#2. Call $(3, 2)$ as M.
#3. $|\overrightarrow{PA} + \overrightarrow{PB}|$ is minimized if O, P, and $(3, 2)$ are collinear.
#4. Then, $PM = \sqrt{13} - 1$. Hence, $|\overrightarrow{PA} + \overrightarrow{PB}|$ has its minimum at $2\sqrt{13} - 2$.

24

#1. $\mathbf{OH} = \mathbf{OA} + \mathbf{OB} + \mathbf{OC}$ implies that $\mathbf{OH} - \mathbf{OA} = \mathbf{OB} + \mathbf{OC}$.
#2. $\overrightarrow{AH} = \mathbf{OB} + \mathbf{OC}$.
#3.

$$
\begin{aligned}
AH^2 &= \|\mathbf{OB} + \mathbf{OC}\|^2 \\
&= (\mathbf{OB} + \mathbf{OC}) \cdot (\mathbf{OB} + \mathbf{OC}) \\
&= \mathbf{OB} \cdot \mathbf{OB} + 2\mathbf{OB} \cdot \mathbf{OC} + \mathbf{OC} \cdot \mathbf{OC} \\
&= R^2 + 2(R^2 - \frac{a^2}{2}) + R^2 \\
&= 4R^2 - a^2
\end{aligned}
$$

Here, $\mathbf{OB} \cdot \mathbf{OC} = \|\mathbf{OB}\|\|\mathbf{OC}\|\cos(2A) = R^2(\cos(2A)) = R^2(1 - 2\sin^2(A)) = R^2(1 - \frac{a^2}{2R^2})$.
#4. Since $AO = AH$, we get $R^2 = 4R^2 - a^2$. Hence, $3R^2 = a^2$.
#5. Since $a/\sin(A) = 2R$, we get $\sin(A) = \sqrt{3}/2$. Hence, $A = 60°$, $120°$. Since A is acute, $A = 60°$.

25 $3a + 5b = 0$ implies that $3a = -5b$. Hence, $\dfrac{a}{b} = -\dfrac{5}{3}$.

26 $(i + i^2 + i^3 + i^4) \times 505 + i^{2021} = i$.

27

#1. Let $z = a + bi$.
#2. $z^2 = (a^2 - b^2) + 2abi$
#3. $|z|^2 = (2 + 2i) - (a^2 - b^2 + 2abi)$ implies that $ab = 1$ and $2 - a^2 + b^2 = a^2 + b^2$.
#4. Therefore, $(a, b) = (\pm 1, \pm 1)$. Hence, $|z|^2 = 2$.

28

#1. Let $|u| = |v| = |u + v| = k$, for some real k.
#2. $|u + v|^2 = (u + v)(\overline{u} + \overline{v}) = k^2$.
#3. $u\overline{u} + u\overline{v} + v\overline{u} + v\overline{v} = k^2$.
#4. $k^2 + u\overline{v} + v\overline{u} + k^2 = k^2$ implies that $u\overline{v} + v\overline{u} = -k^2$.

#5.

$$\frac{u}{v} + \frac{v}{u} = \frac{u\bar{v}}{v\bar{v}} + \frac{v\bar{u}}{u\bar{u}}$$
$$= \frac{1}{k^2}(u\bar{v} + v\bar{u})$$
$$= \frac{-k^2}{k^2}$$
$$= -1$$

[29] Since $x^2 - 2x + 5$ is a quadratic polynomial with integer coefficients, then $1 + 2i$ must be the other root, as it is the conjugate of $1 - 2i$.

[30] By factor theorem, $x = i$ and $x = -2i$ are solutions to $4x^3 + ix^2 + 11x - 6i = 0$. The other factor is $4x - 3i$.

[31] Since $(z - 3)^2 = 15 - 8i$, let $z - 3$ be $a + bi$. Then, $\sqrt{15 - 8i} = 4 - i$ or $-4 + i$. Hence, $z - 3 = 4 - i$ or $z - 3 = -4 + i$. Then, $z = 7 - i$ or $z = -1 + i$. The largest $|z|$ comes from $z = 7 - i$.

[32]

(a) $\sqrt{2} + \sqrt{2}i$ (b) $-10\sqrt{2} - 10\sqrt{2}i$ (c) $\frac{5}{2} - \frac{5\sqrt{3}}{2}i$ (d) -3

[33]

(a) $6\text{cis}(90°) = 6i$ (b) $2\text{cis}(90°) = 2i$

[34]

$$(2\text{cis}(45°))^5 = 32\text{cis}(225°)$$
$$= -16\sqrt{2} - 16\sqrt{2}i$$

[35]

(a)

$$(1 + i)^{12} = (\sqrt{2}\text{cis}(45°))^{12}$$
$$= 64\text{cis}(180°)$$
$$= -64$$

(b)

$$(1 - i)^{24} = (\sqrt{2}\text{cis}(-45°))^{24}$$
$$= 2^{12}$$

36

(a) $z = 1, -\dfrac{1}{2} + \dfrac{\sqrt{3}}{2}i, -\dfrac{1}{2} - \dfrac{\sqrt{3}}{2}i$

(b) $z = \pm 1, \pm i$

(c) $z = 1, \dfrac{1}{2} + \dfrac{\sqrt{3}}{2}i, -\dfrac{1}{2} + \dfrac{\sqrt{3}}{2}i, -1, -\dfrac{1}{2} - \dfrac{\sqrt{3}}{2}i, -\dfrac{1}{2} + \dfrac{\sqrt{3}}{2}i$

(d) $z = 2, -1 + \sqrt{3}i, -1 - \sqrt{3}i$

37

$$\begin{aligned}
\sum_{n=0}^{\infty} \frac{\cos(n\theta)}{2^n} &= \sum_{n=0}^{\infty} \left(\frac{e^{in\theta} + e^{-in\theta}}{2^{n+1}} \right) \\
&= \frac{1}{2} \sum_{n=0}^{\infty} \left(\frac{e^{i\theta}}{2} \right)^n + \frac{1}{2} \sum_{n=0}^{\infty} \left(\frac{e^{-i\theta}}{2} \right)^n \\
&= \frac{1}{2} \times \frac{1}{1 - (e^{i\theta}/2)} + \frac{1}{2} \times \frac{1}{1 - (e^{-i\theta}/2)} \\
&= \frac{1}{2 - e^{i\theta}} + \frac{1}{2 - e^{-i\theta}} \\
&= \frac{4 - (e^{i\theta} + e^{-i\theta})}{4 - 2(e^{i\theta} + e^{-i\theta}) + 1} \\
&= \frac{4 - 4/7}{5 - 8/7} \\
&= \frac{28 - 4}{35 - 8} \\
&= \frac{24}{27} \\
&= \frac{8}{9}
\end{aligned}$$

38

$$\begin{aligned}
\cos^4(x) &= \left(\frac{e^{i\theta} + e^{-i\theta}}{2} \right)^4 \\
&= \frac{e^{4i\theta} + 4e^{2i\theta} + 6 + 4e^{-2i\theta} + e^{-4i\theta}}{16} \\
&= \frac{1}{8} \cos(4x) + \frac{1}{2} \cos(2x) + \frac{3}{8}
\end{aligned}$$

Hence, $a + b + c = 1$.

Solutions for Topic 8

1

(a)

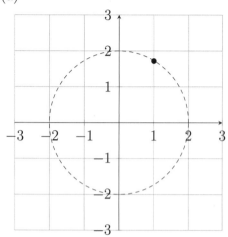

(b)

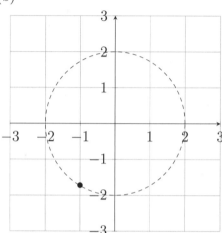

(c)

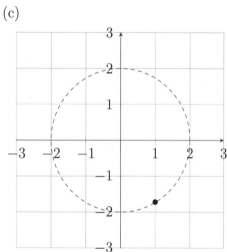

(d)

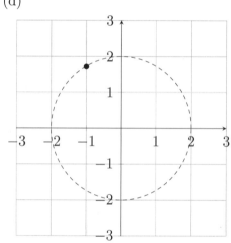

2

(a) $(3\cos(\frac{\pi}{4}), 3\sin(\frac{\pi}{4})) = (\frac{3\sqrt{2}}{2}, \frac{3\sqrt{2}}{2})$

(b) $(-2\cos(-\frac{2\pi}{3}), -2\sin(-\frac{2\pi}{3})) = (1, \sqrt{3})$

3

(a) $(2, -\frac{\pi}{6})$, or $(2, \frac{11\pi}{6})$

(b) $(4, -\frac{\pi}{4})$ or $(4, \frac{7\pi}{4})$

4

(a) $(2, 2\sqrt{3})$

(b) $(-1, \sqrt{3})$

5

(a) $(2, \frac{7\pi}{3})$

(b) $(-2, \frac{4\pi}{3})$

6

(a) $r = \dfrac{3}{\cos(\theta) + \sin(\theta)}$

(b) $r^2 = \dfrac{4}{\cos(\theta)\sin(\theta)} = \dfrac{8}{\sin(2\theta)}$

7

(a) $x^2 + y^2 = 3x$

(b) $x^2 + y^2 = 4$

8

#1. $r\sin(\theta + \frac{\pi}{6}) = r\cos(\frac{\pi}{3} - \theta) = r\cos(\theta - \frac{\pi}{3}) = 3$.

#2. Hence, the line is perpendicular to $\overline{ON}$ where $N(3, \frac{\pi}{3})$.

#3. The x-intercept is 6 and the y-intercept is $2\sqrt{3}$.

#4. Therefore, $y = -\dfrac{1}{\sqrt{3}}x + 2\sqrt{3}$.

9 Since $-1\sin(0) + (-2)\sin(-\pi) + 2\sin(\pi) = 0$, we conclude that the three points are collinear.

10 According to the rule we found above, let (r, θ) be the polar point on the line that passes through the two polar points. Then,

$$1 \cdot r \cdot \sin(\frac{\pi}{4} - \theta) + 2 \cdot r \cdot \sin(\theta - \frac{\pi}{3}) + 2 \cdot 1 \cdot \sin(\frac{\pi}{3} - \frac{\pi}{4}) = 0$$

Hence, $r = -\dfrac{2\sin(\frac{\pi}{12})}{\sin(\frac{\pi}{4} - \theta) + 2\sin(\theta - \frac{\pi}{3})} = \dfrac{\sqrt{6} - \sqrt{2}}{(2 - \sqrt{2})\sin(\theta) - (2\sqrt{3} - \sqrt{2})\cos(\theta)}.$

11 According to the formula we found above, we get

$$d^2 = 3^2 + 7^2 - 2 \cdot 3 \cdot 7 \cdot \cos(\frac{5\pi}{15})$$
$$= 9 + 49 - 42 \cdot \frac{1}{2}$$
$$= 58 - 21$$
$$= 37$$

where d is the distance between $\left(3, \frac{\pi}{5}\right)$ and $\left(7, \frac{8\pi}{15}\right)$.

12 Using the formula we retrieved, we get $r^2 = 6r\cos(\theta - \frac{\pi}{4})$.

13

(a) $25(x + \frac{6}{5})^2 - 20y^2 = 16$

(b) $49(x + \frac{6}{7})^2 + 112y^2 = 64$

(c) $3(y - 2)^2 - x^2 = 3$

(d) $15(x + \frac{4}{15})^2 - y^2 = \frac{1}{15}$

14 $f(\theta) = \dfrac{2}{1 - \cos(\theta + \frac{\pi}{2})}$.

15 Since $e = \dfrac{c}{a}$ in Cartesian coordinates, we get $e = \dfrac{1}{2}$. Therefore, $x = -3$ is the directrix located leftside of the center since $e = \dfrac{PF}{PD} = \dfrac{1}{2}$, and the leftmost x-intercept is $(-1, 0)$, which implies that $PF = 1$. Thus, $PD = 2$ indicates that D is 2 units left of $P(-1, 0)$, so $D(-3, 0)$ must be on the vertical line. Hence, $x = -3$ is the directrix located left from the center.

16

(a) Since $r = f(-\theta) = f(\theta)$, we conclude that the graph is symmetric about the x-axis.

(b) Since $r = f(\pi - \theta) = f(\theta)$, we conclude that the graph is symmetric about the line $\theta = \dfrac{\pi}{2}$, i.e., y-axis.

17

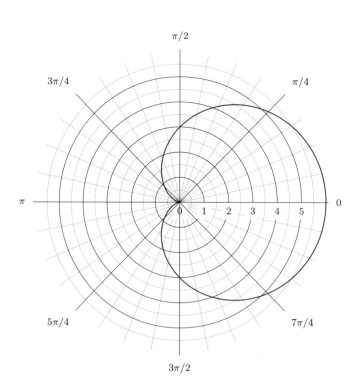

18

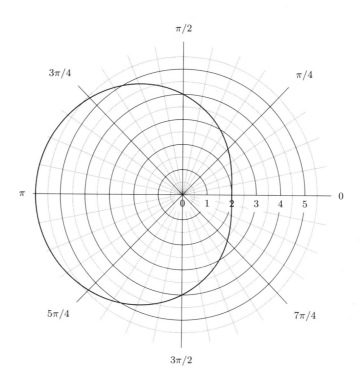

19

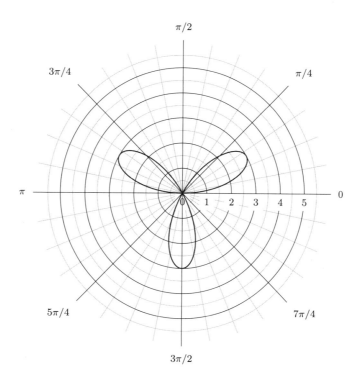

Solutions for Topic 9

1 $\begin{bmatrix} 2 & 3 \\ 3 & -2 \end{bmatrix} \begin{bmatrix} 3 & -1 \\ 2 & 4 \end{bmatrix} = \begin{bmatrix} 12 & 10 \\ 5 & -11 \end{bmatrix}.$

2 $\begin{bmatrix} 3 & 4 & 5 \\ 1 & 2 & 3 \\ 0 & 1 & 2 \end{bmatrix} \begin{bmatrix} 1 & 2 & 3 \\ 2 & 4 & 1 \\ 0 & 4 & 5 \end{bmatrix} = \begin{bmatrix} 11 & 42 & 38 \\ 5 & 22 & 20 \\ 2 & 12 & 11 \end{bmatrix}.$

3

(a)

$$\begin{bmatrix} 2 & 2 & 2 \\ -2 & 3 & 3 \\ 0 & -10 & 0 \end{bmatrix}$$

(b)

$$\begin{bmatrix} 2 & -2 & 2 \\ 2 & 3 & -3 \\ 0 & 10 & 0 \end{bmatrix}$$

(c)

$$\begin{bmatrix} 2 & 2 & 0 \\ -2 & 3 & 10 \\ 2 & -3 & 0 \end{bmatrix}$$

4 $\det \begin{bmatrix} 4 & 1 & 2 \\ 0 & 1 & 3 \\ 2 & 2 & 1 \end{bmatrix} = -18.$

5

(a) $-6p^2 + 8p + 16$

(b) $p = -\dfrac{2}{3}, 2$

6 $\begin{bmatrix} 2/15 & 1/3 & 1/5 \\ -4/15 & 1/3 & -2/5 \\ 7/15 & -1/3 & 1/5 \end{bmatrix}$

7

(a)

$$\begin{bmatrix} d & e & f \\ a & b & c \\ h & i & j \end{bmatrix}$$

(b)

$$\begin{bmatrix} a & b & c \\ 2d & 2e & 2f \\ h & i & j \end{bmatrix}$$

(c)

$$\begin{bmatrix} a & b & c \\ d-a & e-b & f-c \\ h & i & j \end{bmatrix}$$

8

(a)

$$\begin{bmatrix} 6 & -4 & 0 \\ 0 & 8 & 4 \\ 3 & -2 & 0 \end{bmatrix}$$

(b)

$$\begin{bmatrix} 0 & 8 & 4 \\ 6 & -5 & -1 \\ 6 & 3 & 3 \end{bmatrix}$$

9 $\begin{bmatrix} 2 & 1 & 1 & | & 4 \\ 0 & 3 & 4 & | & 7 \\ 0 & 3 & 7 & | & 10 \end{bmatrix}$

10 $\begin{bmatrix} -6 & 0 & 1 & | & -5 \\ 0 & 3 & 4 & | & 7 \\ 0 & 0 & 3 & | & 3 \end{bmatrix}$

11 $\begin{bmatrix} -6 & 0 & 0 & | & -6 \\ 0 & -9 & 0 & | & -9 \\ 0 & 0 & 1 & | & 1 \end{bmatrix}$

12 $\lambda = \dfrac{5 \pm \sqrt{33}}{2}$ by solving $(1 - \lambda)(4 - \lambda) - 6 = 0$.

The Essential Guide to
Precalculus

초판발행 2021년 6월 30일
초판 2쇄 2023년 3월 30일

저자 유하림
발행인 최영민
발행처 헤르몬하우스
주소 경기도 파주시 신촌로 16
전화 031-8071-0088
팩스 031-942-8688
전자우편 hermonh@naver.com
출판등록 2015년 3월 27일
등록번호 제406-2015-31호

ISBN 979-11-91188-41-7 (53410)